ACID RAIN AND EMISSIONS TRADING

Acid Rain and Emissions Trading

Implementing a Market Approach to Pollution Control

ROGER K. RAUFER
STEPHEN L. FELDMAN

Energy Center
University of Pennsylvania

ROWMAN & LITTLEFIELD
Publishers

Portions of this book have appeared in summary form in *Public Utilities Fortnightly* and the *Journal of the Air Pollution Control Association,* and we thank these publishers for permission to reproduce those portions here.

ROWMAN & LITTLEFIELD

Published in the United States of America in 1987
by Rowman & Littlefield, Publishers
(a division of Littlefield, Adams & Company)
81 Adams Drive, Totowa, New Jersey 07512

Library of Congress Cataloging-in-Publication Data

Raufer, Roger K.

Acid rain and emissions trading.

Bibliography: p. 147
Includes index.
1. Acid rain—Environmental aspects—United States.
2. Air pollution control industry—United States.
I. Feldman, Stephen L. II. Title.
TD196.A25R37 1987 363.7′32 87-12908
ISBN 0-8476-7555-6

90 89 88 87
6 5 4 3 2 1

Printed in the United States of America

To the late

HAROLD DICKINSON

University of Edinburgh

Contents

Tables and Figures

TABLES

FIGURES

Glossary

APCD	Air Pollution Control District
BACT	Best Available Control Technology
BTU	British Thermal Unit
CEUM	Coal and Electric Utility Model
CIPS	Central Illinois Public Service Co.
CWPS	Council on Wage and Price Stability
DOE	Department of Energy
EEI	Edison Electric Institute
ELI	Environmental Law Institute
EPA	Environmental Protection Agency
EPRI	Electric Power Research Institute
ERC	Emission Reduction Credit
ESP	Electrostatic Precipitator
ETP	Emissions Trading Program*
FERC	Federal Energy Regulatory Commission
FGD	Flue Gas Desulfurization
GAO	General Accounting Office
GE	General Electric Company
ICF	ICF, Inc.
IGCI	Industrial Gas Cleaning Institute
IH	International Harvester Co.
IRS	Internal Revenue Service
LAER	Lowest Achievable Emission Rate
MW	Megawatts
NAAQS	National Ambient Air Quality Standards
NARUC	National Association of Regulatory Utility Commissioners
NO_x	Nitrogen Oxides
NSPS	New Source Performance Standards
O & M	Operating and Maintenance
OTA	Office of Technology Assessment

* See Table 2.1 for a glossary of ETP terms.

PENELEC	Pennsylvania Electric Company
PG&E	Pacific Gas and Electric
PINDEX	An Air Pollution Index
PSD	Prevention of Significant Deterioration
PUC	Public Utility Commission
PURPA	Public Utility Regulatory Policies Act
RACT	Reasonably Available Control Technology
R&H	Rohm & Haas Co.
RFP	Reasonable Further Progress
SCC	Source Classification Code
SEROP	Sulfur Dioxide Emission Reduction Optimization Program
SIP	State Implementation Plan
SO_2	Sulfur Dioxide
TSP	Total Suspended Particulates
TVA	Tennessee Valley Authority
UARG	Utility Air Regulatory Group
UCC	Uniform Commercial Code
USM	Utility Simulation Model
VOC	Volatile Organic Compounds
WP&L	Wisconsin Power & Light

Acknowledgments

This book had its genesis several years ago in a conversation with David Foster, who was then on the U.S. Environmental Protection Agency's Regulatory Reform Staff. In the years since that conversation, numerous individuals have provided both support and direction, many in ways that they little suspect.

EPA staff members (both present and former) have always been willing to listen to—and criticize—our latest ideas, and have saved us much embarrassment by pointing out facts that should have been obvious to us. In addition to Dave, Michael Levin, John Palmisano, Barry Elman, Rob Brenner, and Ralph "Skip" Luken deserve particular mention. Two EPA staffers, Jeannette Austin and John Jaksch, provided tangible support: they arranged for funding to explore some of the ideas that subsequently formed the basis for this book. Both also served as project officers, and in that role influenced our thinking (and this book) to a considerable degree.

The Energy Center at the University of Pennsylvania offered an environment especially conducive to performing the type of work outlined in this book. Bob Wirtshafter, John Keenan, Bob Amundsen, Gitta Howe, and Sybil McRobie all provided helpful assistance; and it seems likely that without the efforts of two indefatigable Energy Center contributors—James Sawyer and Michael Willingham—this book would never have come to fruition.

We imposed freely on the time and patience of numerous electric utilities and Public Utility Commissions in the midwestern and northeastern portions of the United States, particularly in the state of Illinois. Many suffered repeated phone calls that requested detailed, specific information, and to all we owe appreciation far beyond levels already expressed. Several former colleagues in Illinois, including Jay Norco, Alan Jirik, Art Kleinrath, and Ed Bodmer, were particularly helpful.

Finally, a very special acknowledgment for the support of our families. Our young children were not sure what was going on, but they knew we spent a lot of time at our desks when we could have been playing with them. Our wives knew all too well, but chose to cooperate anyway.

Part I

The Market

1 Introduction

Economists have long been fascinated with market imperfections, and the idea of using a market approach to deal with one such imperfection—pollution externalities—seems particularly satisfying. This fascination has generated a rich theoretical literature over the years, concerned not only with market-based mechanisms but also with other economically efficient approaches to pollution control. These include emission taxes (or fees) and various hybrid tax/subsidy and tax/market systems.

Despite this volume of literature, legislators have displayed little enthusiasm for these ideas. Regulatory officials and those faced with the task of actually implementing the proposed control programs have shown even less, and the beneficiaries—those subject to excessively expensive pollution control requirements—have sometimes shown the least enthusiasm of all.

While practical experience with these economic approaches falls far below what one might expect, we are fortunate in the United States in having developed a market-based system that supplements (but has not replaced) the typical command/control program for air pollution control. The regulatory reform provisions that codify the (albeit limited) market approach are commonly known as the Emissions Trading Program (ETP). This program was designed to deal with localized air pollution problems near emission sources and the attainment of National Ambient Air Quality Standards (NAAQS) for pollutants such as ozone, sulfur dioxide, and particulates. It was not designed, however, to deal with a more recent air pollution concern—acid rain.

The research in this book began with two objectives: (a) to determine how the existing market approach would work for an acid rain program, and (b) to determine whether anything could be done to foster this approach. The book subsequently evolved into three parts, each building on the findings of previous sections.

The first part analyzes the existing market program from a user's viewpoint. This part was suggested not only by academic curiosity, but also by the fact that one of the authors had participated in the development of a private sector brokerage arrangement that subsequently failed. The analysis identifies impediments that affect participants in the market, and helps to explain why the brokerage did not work. It also focuses attention on utility-specific impediments, since electric utilities are expected to shoulder the greatest control requirements under most acid rain control programs.

In our discussions with the legal representatives of a utility trade group, an offhand remark laid the basis for an idea that subsequently became a central tenet of this book—that the leasing, instead of the sale and purchase, of Emission Reduction Credits (ERCs) could encourage use of the market approach.

We were anxious to take our ideas about market impediments and ERC leasing to those who would actually participate in the market-based acid rain control program—the electric utilities. Interviews with electric utilities and their regulators, the Public Utility Commissions (PUCs), form the basis for the second part of this book. Even more important than their reaction to our ideas, however, was listening to the utilities describe their experience with the ETP and their views about potential problems under an acid rain program.

The concerns and issues raised in our discussions with the utilities and the PUCs led us to believe that ERC leasing might eventually play an important role in a market-based acid rain control program. We decided to explore its feasibility in a more rigorous, analytical fashion, and this is outlined in the third part of the book.

While our original intention was to determine whether ERC leasing was feasible, we were unprepared for the strength of the results. The analysis showed not only that it is feasible, but also that emissions trading sufficient to meet the targeted emission levels for acid rain control simply will not occur unless leasing (or a similar mechanism) is employed to allow the utilities to maintain control of their ERCs in future years.

By minimizing the strategic uncertainty associated with the sale of such a novel property right, leasing should overcome many of the implementation problems identified with a market-based mechanism for pollution control. It is certainly not a panacea, however, and those considering implementing such a program must first place market-based systems in their proper perspective.

POLLUTION CONTROL PHILOSOPHIES

Societies have attempted to deal with the problem of environmental degradation far back into history. Alison, for example, notes that

> examination of the legal prohibitions . . . of the more than twenty civilizations that have appeared on the face of this planet, some of

> which have long since vanished, reveals consistent attempts by lawmakers to control . . . environmental degradation.[1]

While the nature of the problem has changed over time, the methods used to deal with it remain essentially the same. The obvious method to regulate the problem is for society to order the polluter to stop, and then to punish him if he does not. This approach is more formally known as the command/control approach and has been used for a very long time. For example, a Londoner was executed in 1308 for violating the king's decree banning the use of coal in furnaces within that city.[2] Penalties today are somewhat less severe, but the basic philosophy still applies. Society's "commands" are typically given in the form of emission limitations or standards, and the punishment is expressed as fines or, in severe cases, imprisonment.

The principal advantage of the command/control approach is that it is relatively easy to implement, since compliance can be readily determined. Years ago, those operating odorous processes (for example, tanning facilities and production of printing ink vehicles) were required to move outside the city limits—an easy requirement to enforce. More recently, the presence of black smoke curling from a stack indicated a violation. While the emphasis today has shifted from visual and olfactory detection to continuous emissions monitoring, to identify offenders remains a relatively straightforward task.[3]

While command/control regulations can thus be quite effective, they often require that societies spend more resources to achieve a certain level of environmental quality since these regulations usually ignore differences in the marginal costs of control between different types of emission sources. These cost differences can be substantial.

To address these cost differences, economists have proposed two alternative approaches to command/control regulation: (a) emission taxes and (b) a market approach, based on some type of pollution rights. Both approaches introduce an element of freedom of choice for the regulated emission source; through self-interest, the source is expected to conform with the overall social goal of economic efficiency.

The most widely discussed economic approach for the treatment of pollution externalities is a tax on emissions or effluent. First discussed in the 1920s by A. C. Pigou in his classic work *The Economics of Welfare*, the idea has generated an extensive theoretical literature.[4] In an ideal world, a marginal damage function for a pollutant could be determined, along with a marginal cost function for its control. The tax would be set at the point where the marginal damage equaled the marginal control costs. Each emission source then would install pollution control equipment until it reached the emissions level where its marginal control cost equaled the emissions tax; after that point, it would pay the tax rather than install equipment.

While the literature on market systems is not as extensive, considerable attention has also been focused on this approach during the past 20 years. J. H. Dales's *Pollution, Property and Prices* (1968) outlined in a modern context a concept that has been termed "a disarmingly simple and powerful one . . . Its intellectual genesis is to be found in the realization that the behavioral source of the pollution problem [can] be traced to an ill-defined set of property rights."[5]

The market approach formally defines the right to emit pollutants, provides an allocation scheme for these rights, and then allows those who have the rights to exchange them in a market. A variety of market approaches have been suggested, with most addressing pollutant emissions within an airshed (or at least those from major sources); allocation schemes are usually based on either an auction or some type of grandfathering system to minimize wealth or income transfer. As one would expect, the element of freedom introduced by these two approaches also has its price. Economic approaches are not easy to implement, and this in large part explains the reluctance of regulators to adopt these types of programs.

We will discuss the relevance of these approaches to the acid rain problem in later sections, with particular attention to technical implementation concerns of the market approach. Throughout this work, economic rationality in environmental public policy decisions is treated as an underlying goal. Issues such as societal attitudes toward polluters, industries' motivations for controlling emissions and the legitimacy of placing a "price" on pollution have, for the most part, been ignored. This is not to say they are unimportant.

NON-ECONOMIC CONCERNS

A certain amount of "baggage" must be carried on the search for economic efficiency, and in many cases the weight of this baggage may make one question the need for the trip.

The first burden is the recognition that economic rationality does not necessarily coincide with political rationality. The rational planners of the 1960s who tried to introduce the Planning, Programming, and Budgeting Systems (PPBS) programs made the mistake of not only ignoring the political process, but of actually denigrating it.[6] They learned very quickly that political costs and benefits are crucial. Costs are associated with every regulatory action, starting with the psychological costs of simply changing the way things are done in the bureaucracy. As we will discuss, the economic approaches can also have substantial wealth transfers, and the political side effects can be very problematical.

Majone has studied this problem and concluded that the performance of policy instruments for pollution control "depends more on the institutional framework within which they are used than on their technical characteristics."[7] He notes that emission taxes could work, but only "in a political system radically different from the one we know."[8]

It is perhaps appropriate that a market system has evolved in the United States. The basic values of the social system in the United States are the sanctity of private property, limited government, and individual liberty—all of which are consistent with the creation of property rights and increased freedom of choice for the emission source found in the market system.

But the capitalist's reliance on the market is, of course, heretical to the socialist. As Sandbach puts it, "market price is not . . . equivalent to any intrinsic value but a distorted value which serves the interests of the ruling class."[9] The themes of maintaining the existing distribution of wealth for the ruling elites and of oppressing the poor by extracting a price for natural amenities (for example, clean air) are found throughout socialist writings on the environment.[10] French socialist Andre Gorz goes even further, claiming that the "ecological perspective is incompatible with the rationality of capitalism."[11] Certainly new groups at the frontiers of the ecological movement, such as Earth First! or the nonanthropocentric "deep ecology" groups, are unlikely market-system supporters.[12]

The belief that it is inappropriate to use a pricing mechanism for pollution control is not limited to radical socialists or environmentalists. Many American citizens believe that industry should not be provided with a "license to pollute"; they feel that industry pollutes with the sufferance of society—certainly not with its blessing. A market mechanism would serve to legitimize the pollution.

Political scientist Aaron Wildavsky and anthropologist Mary Douglas have produced some very perceptive thinking about these beliefs and about related "non-economic rationality" issues concerning the environment.[13] Their research about the moral and social components of impurities, the ritual of cleansing, and the cultural views about risks from technology and pollution provides a starting point from which to address issues that cannot be ignored if economic approaches are ever to play a major role in the future.

A MAJOR ROLE IN THE FUTURE?

Considering the poor reception that legislators and regulatory officials have given to the economic approaches and the non-economic concerns described above, can we see a future for these mechanisms?

Some view the development of emissions trading as a first step in the transition from command/control regulations to the economic approach.[14] Two Clean Air Act observers have even suggested that the latter mechanisms could predominate by the turn of the century.[15] We are not as sanguine. We also believe that the market mechanism has a definite role to play in the future of the air pollution regulatory process in the United States, but recognize that the role will continue to be limited.

Two major issues are at stake here. The first is whether economic mechanisms can legitimately be applied to an environmental system that does not necessarily encompass a common paradigm or worldview. The second concerns the ability of the regulatory infrastructure to support the market approach.

Environmentalists have argued successfully in the past that the costs of controlling pollution should not even be considered in setting the NAAQS, and many of the emissions limitations were determined regardless of cost and technical feasibility. On the other hand, much of Wildavsky's work on non-economic concerns began as a search to find some way to view environmentalists' actions as rational; it certainly was not readily apparent in actions where large amounts of money were spent for pollution control to achieve very little clean-up.[16]

Some type of peaceful coexistence is obviously preferable. The second law of thermodynamics ensures that we will always have waste products. As James Lovelock eloquently stated in *Gaia*, "pollution is not, as we are so often told, a product of moral turpitude. It is an inevitable consequence of life at work."[17]

While the worldviews of economists and environmentalists may differ considerably, environmentalists have begun to realize that they can often use economics to achieve their own goals. For example, the Environmental Defense Fund (EDF) developed an economic model to show that energy conservation and alternative supply sources were preferable to nuclear power plants.[18] Similar efforts have focused on minimizing wastes and recycling.[19]

We view the ETP in a similar way, the use of economics to achieve an environmental goal. We do not propose to use economics to determine the Pareto optimal solution to the emissions reduction problem for acid rain; in real life, this will be solved in the political arena, not by economics. But once the environmental goal has been selected—whether a 2 million, 10 million or 20 million ton reduction in SO_2 emissions—the market approach can help achieve that target in a cost-efficient manner.

The second major issue deals with the ability of the regulatory infrastructure to employ the market mechanism. When our early ancestors polluted a particular environment, their solution to the problem was simple: they moved to a new site. When man stopped moving and congregated in cities, the command/control mechanism became the principal means of pollution control, as our 14th-century Londoner would attest. Unfortunately, a lack of understanding about the impacts of pollution and a lack of technological options for control made these cities grim places from an environmental viewpoint.

Recent use of command/control regulation has been much more effective, and a relatively sophisticated infrastructure has been developed to deal with pollution. This has not come cheaply; the United States has spent more than half a trillion dollars for pollution abatement and control since the early 1970s.[20]

As society climbs the marginal cost curve of pollution control, it becomes more and more expensive to achieve ever smaller levels of pollutant reduction. The inefficiencies of the command/control approach do not seem quite as important at the lower end of the curve, when action is needed and an infrastructure doesn't exist. Indeed, the strength of command/control lies in its ability to get things moving without complication. But once the infrastructure is in place, it seems foolish to ignore the lessons of economics.

Liroff has noted that the debate about emissions trading is based in the differences of philosophy between the "command expansionists" and the "command minimalists."[21] The command expansionists contend that the infrastructure isn't good enough; that the State Implementation Plans (SIPs), emission inventories, and other parts of the infrastructure are too unreliable, technically incomplete, and too easily manipulated by industries to allow economic mechanisms to be employed readily. The command minimalists, on the other hand, point to the considerable cost inefficiencies in command/control regulations, and contend that the market mechanism can be—and already has been—used to identify and shore up weaknesses in the existing command/control program.

We are firmly in the command minimalist camp. The concern about emissions inventories and related data is a legitimate one, but it is not clear that any data base could ever hope to reach the level of accuracy and detail that expansionists would consider adequate, particularly given the dynamic pattern of emissions and the limited resources and funding available to develop and maintain the data base.

While the projection of attainment of the NAAQS is perhaps the most critical part of the State Implementation Plan (SIP), it is important to realize how this is usually formulated. State agencies rely on a variety of data bases (such as emissions, meteorology, monitored air quality), that are often of suspect quality. These are then used as input for standardized Gaussian dispersion models or other analytical techniques that usually give only a rough approximation of real conditions.* With a tight deadline approaching, the individual states are expected to take the results, apply various command/control requirements such as Reasonably Available Control Technology (RACT), and then develop an approach to achieving the standard that bears some semblance of economic rationality. Liroff cites one of the command expansionists as saying, "It is necessary to understand how tenuous—how 'fictional'—SIPs really are."[22]

Some SIP fictions are deliberate. For example, emission sources with a stack height higher than Good Engineering Practice consume Prevention of Significant Deterioration (PSD) increments at the theoretical level, as opposed to the actual pollution. States use dispersion models to track these increases, since they do not actually exist and in any event would be virtually impossible to detect in the noise level of the monitoring data.

Most of the fictions are not deliberate, however, but are usually the result of too little time, not enough money and manpower, or simply too many unknowns. We do not wish to denigrate the work of the state control agencies, which are doing a good job given the constraints under which they must perform. Unfortunately, however, in the real world the data will never be good enough and in any event are changing all the time.

SIPs must be recognized for what they are—constructs that provide a general framework for controlling air pollution. The SIP technicians at the state agency will never have as much information as the plant engineers. Nor should they want it. The regulator can maintain control over transactions in a market, and this is where individual actions can be fine-tuned. As long as the general framework is sound, environmental goals will be achieved. Even if one believes (which we don't) that the only way to upgrade the SIPs is to collect better emissions data from point sources, it seems more likely that the information will be obtained through market incentives rather than through further browbeating under command/control.

In the United States, the market approach that has developed is a constrained one. It builds on the general framework of the SIPs and command/control requirements, supplementing these programs instead of replacing them. The program takes considerable pains to ensure that the non-economic concerns are addressed. It doesn't trade "rights to pollute," but rather Emission Reduction Credits (ERCs), which are surplus emission reductions (that is, pollution control beyond levels

* Short-term dispersion models are considered good if they predict concentrations within a factor of two or three; many states developed SIPs using the rollback model, which essentially proportions emissions and air quality concentrations to determine projected compliance levels. For ozone, Appendix J or EKMA are relatively simple approaches, given the complex photochemistry they must represent.

required by the command/control regulations). Liroff has noted that "regulators do not want to create too concrete a 'property right' in emission credits."[23]

Although the market in the United States has been constrained, it nonetheless represents the most advanced trial of such a market mechanism in the world. Its very existence is due to the sophisticated infrastructure that serves as its base, and the capitalist milieu in the country has not hindered its development. The difficulty in implementing the program even under these favorable conditions, however, does not bode well for its use in other countries. Certainly, developing countries are most in need of economic efficiency, particularly since funds spent on pollution control equipment are often viewed as being non-productive. Many of these countries have never even taken the first steps in a command/control program, however, and the potential for implementing a relatively complex market system seems remote at best. The probable thinness of markets, government control of major pollution sources, and endemic corruption among regulators could cause implementation problems even if a willingness to experiment existed.

The future role of the market mechanism will probably continue to be limited by command/control and non-economic concerns, and our intuition tells us that some of these limits are not necessarily bad. Emissions trading offers a real world opportunity to minimize pollution control costs, however. The analysis of market impediments and the proposed ERC leasing mechanism outlined in this book are designed to stimulate the market approach, keeping within the boundaries of the existing constrained program (or stretching them slightly in a few cases). We hope that it will enhance the existing air pollution control program in the United States and ultimately lead to a more economically efficient program to mitigate acid rain.

2

Overview of Control Strategies for Acid Rain

THE ACID RAIN PROBLEM

"Acid rain" is a powerful phrase that has recently entered the general public's lexicon, and editorial cartoons showing persons standing under an umbrella with only the spokes remaining have become common throughout the print media. Increased public concern has resulted in increased funding for scientific research, and several thousand national and international research projects have recently been completed or are currently underway.[1] This has enabled research on the issue to move from mere problem identification toward a search for understanding—of sources of relevant emissions, atmospheric transport, air, soil, and water chemistry, resulting environmental impacts, as well as the likely impacts of various control strategies.

Precipitation is naturally acidic. The background acidity arises from atmospheric carbon dioxide, sulfur and nitrogen compounds from natural sources, and weak organic acids (formic and acetic).[2] At the South Pole and certain other remote sites (for instance, islands in the Pacific), the amount of acid deposited is relatively small and usually includes significant contributions from weak organic acids rather than from the strong acids (sulfuric and nitric) associated with fossil fuel emissions.

It is clear that man's activities have altered the chemical composition of precipitation in eastern North America from its pre-industrial values, for major industrialized regions experience acidity in surface water and groundwater far in excess of that from natural sources. The highest levels of acid deposition are in the regions with the highest levels of man-made sulfur and nitrogen emissions. Man-made emissions of sulfur, nitrogen, and other acid precursors in eastern North America predominate (greater than 90%), compared to the natural emissions of these substances.[3] The major substances responsible for acid deposition from anthropogenic sources include wet and dry deposited sulfates, sulfur dioxide, nitrogen oxides, nitric acid, trace particulate nitrates, and hydrogen chloride.

Significant fractions of these precursors are transformed to acidic materials and deposited on the surface before reaching the eastern boundary of the continent. Both SO_2 and NO_x react in the atmosphere to produce compounds that make water acidic. The increasing acidity of groundwater comes from both wet and dry acid deposition. In true "acid rain," the sulfur and nitrogen oxides are absorbed by non-gaseous water in the atmosphere and deposited directly on the ground or in natural bodies of water; thus the term wet deposition. Wet deposition includes any of the numerous forms of precipitation, such as rain, snow, hail, and fog. Dry deposition consists of gases and solid particles that settle to the surface of the ground with the potential for acidifying on contact with surface moisture. The dry deposition is often absorbed by surface water (typically after precipitation occurs), and thus also ends up in natural surface water bodies or groundwater.[4]

The atmosphere transports many air pollutants hundreds of miles before returning them to the earth's surface. During this long-distance transport, the atmosphere acts as a complex chemical reactor, transforming the pollutants as they interact with other substances, moisture, and energy. Under the right conditions, emissions of sulfur dioxide and nitrogen oxides are chemically transformed into sulfuric and nitric acids during transport. In eastern North America and northern Europe, the total atmospheric acidity is due to sulfur and nitrogen in approximately a 1.5:1 ratio. The ratio varies considerably with season and region, and in some areas, such as the western United States, it can be closer to 1:1.[5]

Predicting the exact surface location at which a given emitted molecule will produce acidification is not possible. The vagaries of weather preclude the prediction of anything except statistical expectations of the location at which the molecule will return to earth.[6] Gaseous sulfur can remain suspended in the atmosphere for weeks before being deposited. Time allows the winds to carry sulfur and its end products long distances from their origin; a small fraction of the sulfur originating in North America is even deposited in Europe. In eastern North America and Europe, roughly one-third of deposited sulfur compounds come from sources more than 300 miles away, one-third from sources 120 to 300 miles away, and the remaining one-third from sources within 120 miles of the deposition site.

Typically, the dry forms of deposition (gases and particles) are found nearer their sources, and wet deposition occurs at longer distances. Estimates of dry deposition are uncertain because no satisfactory technique has been developed to measure it directly. Nevertheless, the available data suggest that in some regions—with

and the implications of a 15 cent / lb. SO_2 tax were evaluated in an early EPA/DOE cost analysis.[18] The latter is reminiscent of the proposed Pure Air Tax of 1972, which the Nixon administration proposed in February 1972.[19] This would have levied a tax on sulfur emissions to the atmosphere, depending upon air quality in the region the previous year. It was thus addressed at the NAAQS, not to acid rain. The tax would have been in the 10 to 15 cents per pound range of emissions for areas not attaining the NAAQS.

Economists have found that substantial cost savings could accompany the implementation of an emissions tax,[20] but politicians have shied away, for a number of reasons. The first problem is the determination of an appropriate tax. Reliable estimates for the damage costs of a ton of specific pollutant, let alone a marginal damage function, are virtually non-existent. Estimates for these data tend to become value judgments by the analyst. Other tax-setting approaches that focus on more quantifiable criteria, such as achieving an ambient standard or even air pollution indices, are therefore often suggested.[21]

For an acid rain program, the marginal damage function could be very difficult to determine. Crocker and Forster have examined ecological impact data for acid deposition, and suggest that the marginal damage function may be non-convex and that damages may be irreversible.[22] They note that a control program for acid rain "may have strong 'all or nothing' elements: intermediate control measures can lead to burdensome control costs while generating few environmental benefits."[23] The relevance of these specific effects on individual water resources, and the aggregate effect of numerous water bodies and categories of damage, are part of the acid rain debate;[24] it is apparent, however, that the setting of an emission tax would be problematical.

Another, and perhaps more important, problem from a political perspective is the distributional impact of the emission tax approach. As noted above, the emission source will apply control technology until the marginal control cost equals the tax, and will then pay the tax for emissions that remain uncontrolled. In many cases, the taxes paid for uncontrolled emissions can be substantial when compared with the control technology cost. Rosencranz, for example, cites tax revenues equal to 60 percent to 100 percent of the technical control cost for SO_2.[25]

Moreover, these numbers represent substantial sums of money: a 15 cent/lb. tax on SO_2 would collect approximately $1.3 billion in tax revenues alone.[26] Utilities and industrial firms are obviously very reluctant to support this type of wealth transfer, and considerable research has therefore focused on tax/subsidy approaches where the tax revenue funds can be returned in some manner.[27]

The difficulty in adjusting the tax if it does not accomplish its environmental objectives, the potential for firms to pay the tax in the short run while the program is still a novelty (and thus exacerbate the appearance of ineffectiveness), and a variety of other implementation concerns have been discussed in the literature.[28] In view of these problems, the lack of political support, and the lack of experience in using this approach, it seems likely that emissions taxes for acid rain control will remain a mechanism safely harbored in the literature. A tax on electrical generation or fuels might be used to subsidize control costs, however.

The third approach, the market mechanism, has been more favorably received by legislators; even early proposed legislation such as the Mitchell bill included provisions for emissions trading.[29] While emissions trading under an acid deposition control program would differ somewhat from the existing ETP, it nonetheless is important to understand how the existing market program developed and operates. Fortunately, the ETP has been the subject of considerable study from both economic and policy science viewpoints, and two recent works provide a cogent overview from each of these perspectives.[30] While our viewpoint is not always identical with that of Liroff or Tietenberg, they nonetheless provide a context for much of the material contained in later chapters, and we heartily recommend them to the reader.[31] The remainder of this chapter provides a brief overview of the ETP for those unfamiliar with its development; a summary of the principal terms are outlined in Table 2.1.

THE DEVELOPMENT OF AN AIR POLLUTION MARKET

As with many other air pollution issues in the United States, the origin of the market can be traced back to the Clean Air Act of 1970, a landmark in environmental legislation. In order to "protect and enhance" the quality of the nation's air resources, the act followed the command/control philosophy and (a) required the setting of NAAQS for air pollutants to protect the public's health and welfare; (b) allowed the states to determine individually the method of attaining and maintaining the NAAQS through emission limitations, but subjected the resulting SIPs to federal review; and (c) set very tight deadlines for meeting the standards. The act had numerous other provisions, including ambitious limitations on automotive emissions.

When it became apparent that certain (primarily urban) areas would not meet the NAAQS by 1975, the states were faced with a dilemma. Since they could not issue permits for new sources in these nonattainment areas, industrial and economic growth in many major cities would

Table 2.1 - A Glossary of ETP Terms

Bubble Policy: This allows existing sources of air pollution to develop alternate emission control strategies (which must have an equal ambient effect) to those specified in the applicable regulations. Sources can thus imagine their plant to be covered by an invisible bubble and can consider the economics of control in meeting emission reduction requirements.

Emissions Banking: The storage of qualified emission reductions credits for later use in offset, bubble, or netting transactions. A number of locations in the country have established formal emissions banks.

Emission Reduction Credit (ERC): A permanent, quantifiable enforceable emissions reduction beyond the level required by regulations. This is the currency of the emissions trading market approach.

Emissions Trading: A limited market approach to pollution control, including bubbles, offsets, netting, and emissions banking. Formerly refered to as "controlled trading," this group of regulatory reforms is commonly called the emissions trading program (ETP).

Netting: Avoiding new source review requirements by ensuring that the "net" increase in emissions is insignificant.

Offsets: The original component of emissions trading, which allows a firm to construct a new emissions source in a nonattainment area if it secures emissions reductions in excess of the new emission contribution.

be severely restricted. The EPA's answer to this problem was an interpretive policy ruling signed by EPA Administrator Russell Train in December 1976.[32]

This ruling contained a mechanism whereby a new source was allowed to build in a nonattainment area if it would "offset" its emissions by more than a comparable amount. The offset could come from better pollution control or from curtailment of similar emissions in the plant if the source was expanding. Otherwise, the plant might have to install better pollution control equipment on its neighbor, or even buy another plant and shut it down. The offset mechanism was adopted because it solved two immediate problems for the EPA: (a) it allowed major new sources to build in the nonattainment areas, and (b) it did not interfere with (and in fact aided) attainment of the NAAQS. Its impact went far beyond these immediate goals, however, since the mechanism formed the basis for introducing economic incentives into the air pollution regulatory process.

In the past, air has been considered a common good, which anyone was free to use in any manner desired. The problem of the common good has been clearly described by Hardin in his famous essay "The Tragedy of the Commons." When there is no constraint on individual use of a common property, there is no incentive for conservation of the resource. The 1970 Clean Air Act was the equivalent of Hardin limiting the herds on the commons. An existing source was allowed to pollute to a certain extent (that is, it was given a parcel of the assimilative capacity of the atmosphere) as defined in the SIP. The SIP in many cases called for constraint (the source might be called upon to use additional control equipment), but the most important factor was the recognition of emission rights inherent in the parcel. A new source trying to enter the area no longer faced an open common, but rather a limited resource. It was required to "buy" its way into the resource.

While the Clean Air Act of 1970 inherently (and unintentionally) developed the concept of emission rights, the EPA's interpretive offset ruling set up the first market to enable the source to "buy" its way into the resource. The market was primitive. As Yandle has suggested, the market was constrained by the rules that EPA issued as part of its offset ruling.[33] These included the following restrictions:

1. Only new emission sources could enter the market as buyers.
2. The new source was required to use LAER (lowest achievable emission rate) technology for emission control.
3. Offsets had to be for the same pollutant.
4. More than a one-for-one tradeoff of emissions was required.
5. Banking of emission credits was not allowed.

Although the market mechanism in EPA's offset policy had so many constraints that it was almost hidden, it did draw some comments. The Council on Wage and Price Stability (CWPS), for example, pointed out the market-like characteristics of the policy and suggested that the market be improved by allowing the free sale and banking of offsets (that is, lifting restrictions 1 and 5 above).[34]

In the 1977 Clean Air Act Amendments, Congress confirmed EPA's interpretive ruling and made the offset process part of the law. It also began to direct more attention to economic measures in air pollution control. For example, the amendments directed the EPA Administrator, in conjunction with the Council of Economic Advisors, to undertake a study and assessment of economic measures for the control of air pollution. The National Commission on Air Quality, also formed by the 1977 amendments, similarly addressed a number of economic alternatives (including emission fees) in its analyses for Congress.[35]

With congressional interest in economic strategies apparent in the 1977 amendments, the EPA began to move even further in this direction. One of the first changes concerned the prohibition on offset banking. It was argued that an emission source might otherwise be inclined to keep its older, marginal units in operation until it needed the offsets. EPA analyses determined that banking would have both direct and indirect cost savings for business,[36] and in January 1979 the EPA revised its original offset policy to allow banking. The EPA also took into account the other CWPS suggestion and stated that "the State is free to govern ownership, use, sale, and commercial transactions in banked emission offsets as it sees fit."[37] These actions opened the market considerably. EPA issued an Emission Reduction Banking Manual, which contained suggestions for both the design and administration of banking systems.[38]

A number of jurisdictions, such as San Francisco, Louisville, and the Puget Sound District in Washington State, developed formal banking programs. Some states also maintained informal lists of banked offsets. Final banking policy for the states was promulgated in November 1986, more than four and a half years after being proposed.[39] The promulgation of final guidance should encourage the development of this part of the market program.

While the offset provisions described above focus on new sources of pollution, many argued that comparable measures should be available to existing sources. In December 1979 the US EPA published a policy statement in the Federal Register on alternative emission reduction options within SIPs, commonly referred to as the "bubble concept."[40] The bubble concept got its name because it treats emissions as though the plant site were under a bubble; as long as the total quantity of emissions and the air quality impact from the bubble meet SIP requirements, the actual stack-by-stack emissions can be controlled according to economics (that is, those sources with the lowest marginal cost of control can be used to provide the emission reductions). The emission reductions can be obtained from the same plant or even another emission source.

Bubbles are applicable to sources in both attainment and nonattainment areas, with litigation concerning the latter type of bubble having been resolved by the Supreme Court.[41] There are atmospheric dispersion modeling and other technical constraints, but the bubble concept nevertheless represents a major step toward an alternative economic approach.

The currency in this market is the Emission Reduction Credit (ERC). An ERC is an emission reduction that is quantifiable, legally enforceable, and permanent. Perhaps its most important characteristic, however, is that it must be "surplus"—that is, it must go beyond the levels of control already required by the command/control regulations. ERCs can be used in offset, bubble, or netting transactions.[42]

The only comprehensive analysis of the offset program, by Vivian and Hall, is now somewhat dated, since it was undertaken in March 1981. Vivian and Hall noted that until then, approximately 1000 applications for permits using offsets had been made, and approximately 800 permits had been granted. The EPA does not record offsets, but estimates that there were approximately 2500 cases by 1984 (either adopted or being considered).[43] By January 1986, EPA had approved or proposed to approve 50 bubbles, with 34 others approved by states under generic bubble rules (see next chapter). At least 170 others were under development or review.[44] The savings from EPA-approved bubbles were approximately $300 million, and industry estimates for the potential savings from all of the approved or proposed bubbles were approximately $1 billion.

Yet many states have experienced market thinness. The Vivian and Hall analysis found only 32 approved external offset transactions (that is, inter- rather than intra-firm), with 48 total offsets transferred. The majority of these were donated either by another firm or by the state.[45] The authors were able to document prices in only eight cases, ranging from $8 per ton/year for particulates in a Maryland transaction, to $1350 per ton/year for hydrocarbons in a California case.[46] They identified a number of brokers, but noted that no brokers had "risked their own capital to purchase credit for speculative resale."[47] The finding that the vast majority of transactions were internal is important, because economists believe that greater cost savings will occur with external transactions.

3

Market Impediments for Electric Utilities

Since trades in an acid deposition program are likely to be affected by the utilities' experience in the ETP, the literature was reviewed from an electric utility's perspective to determine whether utility-specific problems have already been identified. A number of publications noted the Narragansett SO_2 bubble, the role of Southern California Edison in the Sohio pipeline case, and other utility cases, but there was not much data in the literature.[1]

This is probably not surprising, given the nature of the market to date. While numerous SO_2 transactions have occurred, the greatest market activity has been related to volatile organic compound (VOC) emissions, particularly those associated with surface coating, and to a lesser extent, total suspended particulates (TSP).[2] The introduction of acid deposition control requirements would drastically change the supply and demand of SO_2 ERCs and the role of utilities in the market; to date, however, other industrial sectors have played a more important role.

The market program has been evolving, but at a much slower pace than many had hoped. More than 95 percent of the transactions have been internal.[3] More significant cost savings should be associated with external trades because of diversified marginal control costs across industries, but these have not occurred to any significant extent. For example, a private sector brokerage service developed by a midwestern industry group, the Illinois State Chamber of Commerce, made extensive efforts through mailings, newsletters, a telephone hotline, a free seminar, media coverage, etc., to attract participation of the more than 8000 industry members in that state. Only a handful joined the program.[4] The following sections examine some reasons for the slow development of the market, focusing on impediments to ERC transactions that utilities have faced.

GENERAL MARKET IMPEDIMENTS

In most cases, utilities have suffered from problems identical to those of industry. We will begin by examining those problems common to both utility and industrial sources.

Regulatory Constraints

While the Emissions Trading Program introduced the market mechanism into the regulatory process, no one claims that the market even comes close to representing the ideal market of which economists dream.[5] The market is constrained in several senses. Since emissions trading only supplements the command/control approach, sources are still affected by control technology requirements. Thus Lowest Achievable Emission Rate (LAER), Best Available Control Technology (BACT), and New Source Performance Standards (NSPS) provisions are applicable for new sources, and the determination of appropriate technical requirements for control technology often constitutes a major problem inherent in obtaining a permit. For example, the General Accounting Office (GAO) found that a determination of project controls was a major impediment toward completing one utility external offset case that it examined in detail.[6] The potential for some change in this area, such as allowing a NSPS bubble, has been discussed.[7]

A second constraint has been the number of regulatory obstacles that must be cleared to obtain approval for the transaction. In past years, for example, each bubble was considered a SIP revision. Every time a firm intended to implement a bubble, it would be subjected to extensive technical reviews, with both state and federal approvals necessary. Illinois and other states have provisions for economic impact analyses of SIP revisions, and the resulting process could be considerably time-consuming. EPA has been making efforts to streamline the regulatory obstacles through the use of "generic" bubble rules. These remove the double-approval requirements and allow the states to outline plans for reviewing bubbles as part of the SIP. These general procedures are approved once by the federal EPA. Individual plant bubbles are then reviewed by the state alone using these procedures, and the individual bubbles are not made a part of the SIP. This considerably reduces the time necessary for approval. As of November 1986, ten states had approved generic plans for one or more pollutants, and a dozen others were developing such rules.[8]

While the use of generic bubble rules seems to aid utility trades as well as industrial sources, this may not necessarily be the case.

Kontnik, for example, suggested that proposed guidelines for the ETP failed to provide adequate guidance and encouragement for states to adopt generic rules, particularly on sulfur dioxide and particulate trades.[9] This suggested a continued bias in the market toward VOC sources that do not have the dispersion modeling problems associated with SO_2 and TSP.

A more subtle problem is evident in Pennsylvania. That state originally decided that (a) it would not incorporate generic rules for particulates or SO_2 in its SIP, so each bubble for these pollutants would continue to be considered a SIP revision; (b) it would include a generic rule for VOC, but only for two industrial categories—surface coating and graphic arts; and (c) for these two categories, only internal VOC offsets would be allowed under the generic bubble. The state subsequently revised its stand and proposed to allow generic SO_2 and particulate bubbles; one utility spokesman noted that the proposals were so stringent, however, that they would probably not be very useful.[10] The director of Pennsylvania's air pollution agency has stated that EPA has "oversold" the bubble policy.[11] Thus, EPA's efforts to minimize market constraints may not necessarily be adopted on the state level. Similarly, some of the proposed modifications on the state level may be more likely to assist VOC sources than SO_2 or particulates.

Part of the reason why VOCs have worked so well in the ETP is a third regulatory component of the market—the spatial (or locational) constraint. One of the key directives of the 1970 Clean Air Act was the establishment of NAAQS. These air quality standards must be attained and maintained, and the ETP is designed to assist in these goals. The localized impact of a source's emissions therefore limits the area in which ERCs can be developed or obtained. Because VOC impacts are regional in nature, the impacts of this constraint are relatively negligible. For particulate or SO_2 impacts, however, the localized impact is very important. Proposed generic bubble rules, for example, streamline permit conditions only in those cases where tradeoffs occur within a very small distance (within, say 100 or 500 meters). For distances beyond several hundred meters, it becomes more and more difficult to show an equivalent air quality impact. This constraint is not as easily modified; it would require a shift in the manner in which the EPA views attainment of NAAQS (for example, to allow for interpollutant trades as suggested by GAO).[12] The locational constraint on SO_2 emissions should not play as important a role in an acid deposition trading program, since the focus has been on total pollution loading in a region, instead of on localized concentrations controlled by SIP requirements.

The locational problems associated with demonstrating an equivalent air quality impact through dispersion modeling have often been avoided through another technique—the use of trading ratios. A trading ratio is the ratio of emissions reduced to emissions allowed. Thus, for a 1.2:1 ratio, a source that wants to emit 100 tons per year must reduce emissions of that pollutant by 120 tons per year from other sources.

The use of trading ratios has several attractive features. It can considerably reduce dispersion modeling and other transaction costs. The ratio can be set higher to account for uncertainty in the transaction (for example, the impact of different plant operating conditions) or to account for trades between more distant sources. The use of a variable ratio as a planning tool has even been debated in the literature (that is, to attract desired industry or to keep out undesired plants).[13]

There has already been a shift away from modeling in the ETP. The GAO, for example, noted that "The evidence from California suggests that there has been a reliance on an emissions basis with fixed offset ratios as opposed to an air quality modeling basis for determining needed offsets."[14] It is not clear whether the (nonmodeling) bias of VOC sources has been accounted for in this statement.

The success of VOC sources in the market may indeed augur well for utilities, since the acid deposition problem is similarly viewed as regional and there is also an inability to model the impact of an individual point source. EPRI has noted that "the model accuracy necessary to distinguish utility contributions to ambient sulfate levels has not yet been achieved."[15] While developmental work is proceeding in this area, it is apparent that an air quality-based modeling approach would not work; rather, emission trading for acid deposition will require defined trading ratios. The setting of these ratios could be somewhat troublesome, because of questions about the relative contribution of near versus far emission sources. While it may not be difficult to envisage interstate trades between Ohio and Kentucky utilities, the question arises whether trades between Ohio and New Hampshire (or even Ohio and New York) sources should be allowed. The use of relatively high trading ratios may be required in these cases.

The preceding discussion of trading ratios is important because the ratio operates as a fourth market constraint. Since the formation of the market, for example, it has been required that offsets in a nonattainment area be greater than one-for-one. The intention of this requirement is clear: to maintain progress in cleaning up the nonattainment areas. This constitutes a tax on emissions trading, however.[16] It encourages firms to keep ERCs for their own use, not only because they will be readily available if the plant decides to expand but also because a plant expansion is often "taxed" only to the extent that it uses external offsets. A plant expanding with its own ERCs can "net" out of the review, usually using a one-for-one tradeoff.

These types of constraints affect not only individual transactions; they can have a profound impact on the operation of the market. For example, the last two constraints discussed above (locational impediments and the "tax" on external trades) both favor internal offsets. Since the principal function of any brokerage service is to provide information about external ERCs, these constraints have effectively minimized the role of brokers in the ETP.

EPA anticipates that a move to external offsets will occur because control costs vary more widely across industry, and interindustry trades will provide greater savings than will internal ones. Others who have addressed this issue have come to the same conclusion. Russell, in a study of the 11-county Delaware River estuary region, concluded that "in a complex industrial region, the normal play of economic and technological forces will produce the potential for a significant number of trades."[17] The potential for utility (and other emission source) trades will be fostered to the extent that these constraints can be legally minimized.

Uncertainty

As the Council on Wage and Price Stability noted when the offset concept was introduced, "few things are as debilitating to a market as the attenuation of property rights. Thus, if EPA desires to have an effective and efficient market process, the problem of certainty needs to be considered and addressed."[18] Incremental policy changes have led to a continuing state of flux in the Emission Trading Program—just the opposite of conditions necessary to foster the market. Many of these changes were made to minimize the constraints discussed above. Unfortunately, changes can contribute to uncertainty even when they are designed to simplify the process.

Important structural components of the ETP are still not in place. Although the EPA has consistently encouraged the development of banking programs, states were reluctant to do so until the SIP requirements were finalized.[19] GAO found the role of the emissions bank to be very important, recommending that Congressional Committees "should urge EPA to step up its promotion of emission reduction banking. As revealed in our case studies, this institution has the potential to reduce the sizable transaction costs and uncertainties which have beset external offset negotiations."[20] Utilities have faced the same uncertainties as other emission sources. One utility executive is cited as saying, "We just can't be expected to plan when the rules keep changing."[21]

In addition to the changing regulatory framework, uncertainty remains over a number of fundamental issues. The legal status of an ERC has been discussed in a number of articles and documents,[22] and while there appears to be no major problem in this area, substantive case law has yet to be developed. One distinction that must be drawn is that between the ERC and the permit. The ERC is the commodity that is bought and sold; it allows the owner to modify the source's permit. The permit documents are not the commodity.[23] Since internal offsets are essentially permitting arrangements, the distinction is not particularly important in these cases. Yet for external trades, "sources must be able to construct agreements that not only define what they are buying or selling, but also allow them to plan the financial, tax and other consequences of their trades."[24] Kontnik discusses two types of agreements for external trades: an emission reduction exchange and a control program agreement. When Southern California Edison entered into an agreement to install a scrubber on its Alamitos Unit 6, it entered into the second type of agreement. Kontnik notes that these may provide more certainty than the emission reduction transfer type of agreement, but they are site-specific and may create long-term non-transferable legal obligations between the parties. He concludes that "the control program agreements are not well suited for the development of flexible emission trading systems." Thus, it appears that the quest for legal certainty may not necessarily be consistent with development of the market.

One major uncertainty is the fact that both the legal and tax status of ERCs can vary from state to state, and even on a local basis.[25] For example, a number of local jurisdictions (including San Francisco, Louisville, and the Washington State Puget Sound district) have already developed formal banking programs. While these will be subject to their state SIPs, local requirements may differ on items such as the time limitations of an ERC. This could affect the amortization of the ERC for tax purposes. While these legal and tax differences may be a minor irritant in the ETP, they could become a major problem if interstate trades are attempted.

Utilities face a variety of uncertainties related to the quantification, certification, and retention of ERCs.[26] While most of these problems can be resolved through negotiation, substantial differences can occur on control technology requirements, the quantity of offsets required, and even the eligibility of offsets. For example, all three of these issues were heatedly debated in Pacific Gas & Electric's (PG&E) attempt to arrange an external offset at its Protrero plant.[27] After 21 months of negotiation, the project was canceled.

Of particular concern to many firms using the emissions bank are the provisions for confiscation or forfeiture of ERCs if the NAAQS are changed in the future or if air quality levels get worse and the SIP is modified. Partial confiscation to provide a growth allowance or to allow for reasonable further progress (RFP) in attaining standards has also been discussed. Once again, a PG&E case is relevant. After deciding to shut down two coal-fired units, PG&E decided to bank the ERCs. They were issued a banking certificate, but the district took 50 percent of the emissions. The company believed that it was entitled to the full credit, but environmentalists argued that no credit should have been given. Given the considerable uncertainty about the effects of targeted reductions in pollutant loading in an acid deposition control program, the potential for ERC confiscation will be even higher in this type of program.

Finally, there is uncertainty about future regulatory requirements. Firms are reluctant to explore potential ERCs with a regulatory agency if there is even a remote chance that the reduction would become a requirement at some point in the future.[28]

The viewpoint that the future will hold additional regulatory requirements for emissions control and the bias of the ETP toward internal transactions has led to the hoarding of ERCs. Vivian and Hall found this to be true for large firms, particularly energy-related firms.[29] It is not clear that this will change in the near future. As one utility spokesman noted, "If we knew that this would be the last time in the next 30 years that we would be required to control, we'd probably use the market. But it won't be."[30]

The concept of leasing presents a potential method for overcoming impediments such as uncertainty,[31] and is discussed at length in later sections of this book. The use of options on ERCs, whereby a utility obtains regulatory approval for an ERC before purchasing it, is another mechanism that can reduce uncertainty. Only one case of a utility's use of options was described in the ETP literature, however, and the outcome was not very favorable. Pacific Gas and Electric lost $70,000 in option money when the regulatory agency disallowed the offsets it was going to purchase and the project was canceled.[32]

EPA has explored some other ideas to minimize uncertainty, such as the use of indirect offsets.[33] In this type of program, the source pays a fee in lieu of obtaining an offset. The organization collecting the fee then uses collected revenues to purchase offsets. This minimizes the uncertainty for a firm, because there is always recourse if offsets are not available. It formalizes the brokerage process, however, and may introduce problems for the control agency if ERCs cannot be found. In the case of an interstate acid deposition program, the formal public sector brokerage organization would probably be regional in nature, and there would probably be some problems in getting this established.

Transaction Costs

The costs of finding, preparing, and documenting the ERC must be included as transaction costs in the market. Atmospheric dispersion modeling, engineering and/or other technical consultants, such as an emissions broker, can be costly. For example, Pacific Gas & Electric sponsored a study to find offsets in the San Francisco Bay area. The study, which lasted 10 months and cost $56,000, estimated costs and emissions reductions potentially available at more than 200 major sources in the area. Apparently only one source was willing to sell. PG&E expanded its search and was ultimately able to purchase the ill-fated $70,000 option described previously. GAO feels that these transaction costs are somewhat high because they were the first conducted in the area, but these costs are likely to be substantial for utilities, in any event.[34]

Related to this problem is the lack of information concerning market prices. Lacking specific price information concerning the worth of ERCs, some utility representatives may be unable to evaluate the impact of emission trading measures. The manner in which ERC prices are set is important in determining transaction costs. For example, in the PG&E case noted above, the utility would have paid $1.3 million if it had exercised its offset option; comparable emission reductions on its own plants would have cost an additional $19 million.[35] If the premium to potential ERC suppliers had been raised to reflect some of these savings, PG&E might have been able to reduce the search costs. On the other hand, the proposed facility would not have been economically feasible at some higher price.

To minimize transaction costs, it is necessary that sources have information about both the availability and price of ERCs, and this is a function usually performed by brokers. Vivian and Hall have identified a number of brokers, and the EPA has attempted to encourage the practice.[36] The agency held a conference on private sector brokering that attracted approximately 200 persons from around the country, and also issued a handbook including information about ERC financing, tax concerns, and market analyses.[37] Not all brokerage efforts have been successful, however, mainly for the reasons discussed above. Some pollution control agencies have attempted to assist firms in an informal manner by providing computerized listings of potential ERC sources.[38] While a more formal government brokerage role such as that described above for indirect offsets would serve to minimize transaction costs, the question arises whether brokerage should be in the private or public sector. Some industrial sources, feeling hesitant to expose potential control options, have expressed concern about using government brokerage services. Instead they indicated a preference for the confidentiality of a private broker.[39] Since utilities operate in a more regulated setting than most industries, the use of a formalized governmental brokerage in an acid deposition emissions trading program may be acceptable.

The GAO also suggests that allowing the use of interpollutant tradeoffs will reduce search costs. SO_2 and NO_x tradeoffs would appear to be particularly relevant for acid deposition controls. McLean has suggested that an emphasis on SO_2 control is preferable,[40] so the tradeoff ratio might not necessarily be 1:1. Reductions of NO_x to $S0_2$ on a 2:1 basis have already been discussed, but trading in NO_x raises concerns about the emission baselines.[41]

Other Concerns

A number of other factors may have affected both utility and industrial use of the ETP. The demand for an ERC essentially occurs in three situations: (a) when the firm is planning to build a new facility, (b) when it is expanding an existing facility, or (c) when it is required to reduce emissions because of a change in the SIP. Economic growth during the early stages of the ETP was not particularly robust, and the first two demand categories were no doubt affected. Utilities were faced with decreasing projections for electricity demand, and reacted accordingly. The PG&E external offset case (discussed earlier) was withdrawn after revised forecasts indicated lower electrical demand. A PG&E spokesman indicated that the utility had 3600 MW of proposed generation in the planning stages several years earlier, but these were canceled due to the drop in electrical demand. No major baseload units are scheduled to come on line until after the year 2000.[42]

The third category, reduced emissions due to a SIP change, was important after the 1977 Clean Air Act Amendments, particularly for VOC emissions. The impact was somewhat less for TSP and SO_2. For example, many SIP revisions focused on control of industrial fugitive emission sources (for example, storage piles), which could often be controlled in a relatively cost-effective manner compared with previous capital-intensive expenditures for control equipment for stack emissions. This would of course change substantially if an acid deposition control program were instituted.

Another factor was the novelty of the program. GAO found that this contributed to many of the problems identified as impeding market transactions, and noted that this has "behavioral implications which suggest resolution over time."[43]

UTILITY-SPECIFIC CONCERNS

The sections above have discussed impediments that have affected both utility and industrial sources. A number of other items specific only to utilities include rate making, construction planning times, ERCs for customers, and the general regulatory framework.

Rate Making

In addition to tax and regulatory risks, utilities must be aware of rate-making risks before using the ETP. The primary goal of the rate-making body is to provide its customers with the lowest possible cost rates while assuring service reliability, and the use of ERCs may be quite consistent with that objective. The philosophies of attaining this goal may vary from state to state, however, so some risk is involved in the recovery of these costs.[44]

Stathos and Treitman suggest a number of questions that the utility should ask before investing in an ERC, such as, Will the certification costs of an ERC be allowed in the rate base? How will gain or disposition of the ERC be treated? Does it belong to the shareholders, the consumer, or both?

The issue has also been addressed by Hahn and Noll, but in the context of a full scale market trading program, not the limited market associated with ERCs. They note that: "The assumption that electric utilities will minimize costs is debatable . . . If utility regulators are more likely to allow some kinds of costs than others, or are willing to let utilities earn greater profits on some kinds of assets than others, the utility regulatory process will cause the regulated firm to depart from cost-minimizing choices of technology and operating methods."[45] They discuss differences between states using "original-cost" accounting procedures and those using "replacement-cost" in a full-scale emission trading market. They conclude that problems exist in the former that will not occur in the latter. Specifically, the initial allocation of emission rights for the utility have a zero acquisition cost if a "grandfathered" initial allocation mechanism is employed, and hence must be carried on the books at zero value. Any adjustment of a rate of return allowed on the asset will produce a zero profit allowance. Furthermore, they note that most states use the principles in the Uniform System of Accounts established by the Federal Power Commission (the predecessor of FERC), which requires that any gains or losses from the sale of an asset (calculated as the difference between sales price and book value for rate-making purposes) be passed on to ratepayers. This would amount to a 100 percent tax on sales revenue if a utility decided to sell a grandfathered permit. They conclude that "the obvious solution to the problem is to convince original-cost states to adopt replacement cost methods for evaluating emissions permits,"[46] but they don't hold much hope of regulators doing so.

While Hahn and Noll are dealing with a full-scale market trading approach, their work could have some implications for utility ERCs in the ETP. ERCs can be obtained in a number of ways; for example, the scrubber that was to have been installed at Southern California Edison in the Sohio case constituted the full ERC in that case. On the other hand, PG&E banked ERCs created by retiring two coal-fired units. The original cost of newly installed control technology is easily documented, but the original cost of the ERCs created by retiring units presents the same problems described above by Hahn and Noll. Emission rights were given unintentionally through the Clean Air Act, at a zero cost to the utility. Thus all ERCs may not be equal from a utility accounting perspective.

This same point is developed by Gerard, who approaches the problem in a slightly different manner. He makes a distinction between a firm purchasing a banked ERC and one creating an ERC "in the course of an offset transaction where a source's emissions have been reduced below the SIP baseline."[47] In the former, the firm gains an intangible asset (the banked ERC), while in the latter it gains a tangible asset (the pollution control equipment). He then notes that these two ERCs "are not likely to be perfect after-tax substitutes for each other."[48] The decision whether to purchase an ERC or to install pollution control equipment in an acid deposition control program may be affected by these differences.

Construction Planning Times

One factor that differentiates utilities from industrial plants is the relatively long lead times from the proposal stage to actual plant operation. Utility lead times are in the six-to-twelve-year range and have been increasing in recent years.[49] These long planning times may be incompatible with many of the proposed emission banking programs that want to expire or depreciate the value of ERCs held over a long period in order to foster economic growth.[50] Some emission banks have limited lifetimes of five and eight years in certain locations, while others allow permanent lifetimes with no limit.[51] This problem was specifically considered by California, where districts are authorized to establish "an emission offset bank for a specific power plant at a specific location." The provision was "intended to provide some degree of certainty for utilities in finding adequate offsets for power plants early in the siting process."[52]

Long lead time could affect an acid deposition program in a number of ways. First, if utilities feel that existing plant emissions that are retired will be discounted in an existing bank (see, for example, PG&E's 50 percent discount on two coal-fired units) or depreciated over time, they will keep marginal units in operation until reductions are mandated and their full value can be realized. Second, the ten year compliance period in a number of proposed congressional acid deposition bills may conflict with the shorter ERC lifetimes in certain banks. Finally, while conflicts between utility planning requirements for new sources and limitations on ERC life will probably not be a major impediment in many cases, it could occur at certain sites.

ERCs for Customers

While most firms have tended to hoard ERCs for the reasons discussed above, utilities have an additional reason to provide for growth allowances for future customers. Beck has suggested that this is a particularly important factor from the utility's viewpoint, and is likely to lead to a continuation of ERC hoarding even in the face of a major control requirement.[53] The issue is important because it takes into account the future economic growth of an area, and is therefore analogous in many ways to the Prevention of Significant Deterioration (PSD) increment. States are quite protective of their increment, and the experience of utilities facing interstate increment consumption has been of conflict rather than of interstate agreement. If growth becomes an issue in both the air pollution agency and PUC determinations, the interstate trading of ERCs could become quite rare. The manner in which new power plants are introduced into the acid deposition market would also be a factor affecting the utility's desire to hold ERCs.

Potential for Future Regulatory Actions

While the desire to hold ERCs in case of future regulatory actions is probably common to all industries, for utilities it is even more likely. The likelihood of control requirements for acid deposition has been increasing over the past several years, and that alone would be sufficient to hinder the sale of any SO_2 or NO_x ERCs generated by utilities in the past few years. The utilities would no doubt agree with Crandall, who argues that regulators have "an obvious incentive to bury regulatory costs in an industry which can absorb them without immediate employment or output effects."[54] He goes on to state that "the costs imposed upon utilities for pollution control is not proportional to their importance as polluters. In many areas they pay far more than their share of control costs."[55] The magnitude of utility emissions and the potential to affect changes through PUCs have no doubt made them obvious regulatory targets. Utilities do not foresee this changing in the near future.

The potential use of ERCs may also be affected by other regulatory actions, such as the Public Utility Regulatory Policies Act of 1978 (PURPA). For example, where a potential industrial customer may also wish to be a cogenerator (under the terms of PURPA) but lacks required ERCs, "depending on the utility's disposition towards the cogeneration project, an ERC, or the need for it, could be used as an incentive or a deterrent to the success of the cogeneration project."[56] It is unlikely that ERCs could be used very often to defeat the objectives of PURPA, but the underlying point is important. The nature of energy supply is changing.

As noted earlier, PG&E does not have any major baseload units coming on line, but it does have new capacity scheduled. The new capacity is based on renewable resources, geothermal units, and cogeneration.[57] ERCs may play an important role in these energy supply changes. For example, a number of publications have considered ERC creation by solar energy and energy conservation.[58] Minnesota has made provisions whereby district heating systems can add new users by shutting down the client's existing boilers, with offsets originating in the shutdown.[59] California has authorized a special emissions bank, with extra features, for alternative energy sources including cogeneration plants, waste fueled plants and biomass-fueled plants;[60] California formerly placed the onus for ERCs for cogeneration units on the growth allowance of the pollution control agency.[61]

In an acid deposition control program, the effect of energy supply alternatives may become even more important. One Wisconsin utility spokesman has indicated that the large costs associated with an acid deposition control bill could make the utilities noncompetitive with self-generators.[62] This may suggest that utilities would pay more attention to emissions trading to minimize their avoided costs. Utilities are in a business that has been subject to considerable pressure and change in recent years, and they expect that to continue into the future.

4

Other Concerns in the Market for Acid Rain Controls

The utility industry is currently in a period of considerable flux. There have been recent calls for deregulation (to separate electrical generation from transmission and distribution), and the use of electric energy brokering, where utilities negotiate energy transactions on an hourly basis, has increased.[1] Many of these changes are in the general direction of using free market principles to increase utility efficiency, despite the extensive regulatory structure that exists. Emissions trading fits comfortably within this general movement.

Previous sections have described the problems of electric utilities within the existing market framework, and the implications for an acid deposition control program. This section examines some of the other issues that will define emissions trading in an acid deposition context, including the potential for interstate trades, and economic and organizational issues concerning electric utilities that will play a role in implementation.

Numerous cost impact analyses addressing various pollution control strategies have been performed, and many of these have included assumptions for intra- or interstate trading of emissions. ICF summarized the results of several of these analyses and found that interstate (rather than intrastate) trading provisions reduced annualized costs by $400 to $900 million.[2] The Edison Electric Institute (EEI), however, assumed no interstate trading in its analysis, noting that it is "totally inconsistent with political reality."[3]

The GAO conducted an analysis of one bill (S-3041) that was approved by the Senate Committee on Environment and Public Works in the 97th Congress in August 1982. The bill would have allowed emissions trading, but the GAO analysis gave credence to EEI's position. The report noted that considerable controversy surrounds the question of emissions trading, given the relative lack of experience with the concept. It concluded: "Our examination of the issue suggests that significant cost savings can result from emissions trading *within*

States; but that institutional constraints reduce the potential for *interstate trading*" (emphasis in original).[4]

A more detailed cost analysis of S-3041 conducted by ICF, Inc., suggests that the savings associated with interstate (rather than intrastate) emissions trading through 1995 are even lower than the figures cited above.[5] Savings were determined to be approximately $200 million. This was attributed to the fact that the allocation scheme in S-3041, which is based on emissions exceeding a 1.5 pounds SO_2 per million BTU heat input, quite closely follows the economics of control. In ensuing years, however, the savings become increasingly important as trading plays a more critical role in new source growth.

In the GAO analysis of S-3041, a number of Public Utility Commissioners were contacted, although the survey was not comprehensive.[6] The response was not very favorable for interstate emissions trading, and the executive director of one New England PUC is quoted as saying: "We're already paying as much for electricity as anybody, and the thought of paying for a utility in another state to reduce its emissions doesn't make sense."[7]

The debate about acid deposition is seen as a regional conflict, and the PUC commissioners obviously hold various viewpoints. The GAO notes that their "interviews and literature review on interstate trading's feasibility range from uncertainty—to blatant skepticism."[8] These negative findings about interstate trading should perhaps be accorded even heavier weighting, given GAO's earlier enthusiastic endorsement of the market approach to emissions control.[9]

The historical record on interstate trading in the ETP is very sparse. Vivian and Hall discussed only two attempts at interstate emissions trading, neither involving an electric utility.[10] In the first case, "Pennsylvania authorized a permit for a new source in which an offset was to be obtained from a plant 75 miles away, in New Jersey." The project was canceled when the firm changed its production plans, but the report notes that the "interstate transfer of offset credit was approved by U.S. EPA Region III."[11] Similarly, they discuss a permit approved in South Carolina that used part of a banked credit from the shutdown of a plant in North Carolina.[12] In both cases, perhaps the most important factor was that a single firm owned the plants in both states.

Al Courtney, formerly of Commonwealth Edison Company and Chairman of the Utility Air Regulatory Group's (UARG) Acid Deposition Committee, believes that a comparable arrangement would be the only possible situation in which an interstate trade would be feasible. He noted that utilities such as American Electric Power, the Southern Company, and TVA might be able to arrange interstate trades of internal ERCs, but only after detailed negotiation with the states' PUCs. He believed that external ERC trades are unlikely, however, even within state boundaries, since "it places the utility's well-being in someone else's hands."[13]

The Environmental Law Institute (ELI) addressed the Institutional Aspects of Transported Pollutants for the National Commission on Air Quality, and found that "states opting for the market-based mechanisms will have to invest substantial resources in designing these programs and getting the necessary statutory and regulatory authority."[14] The analysis considered only intrastate trading, and estimated that three to six years would be needed for the development of state regulations and an additional one to four years for compliance. While ELI noted that the state's efforts would benefit from the existing ETP, the states "will encounter many novel implementation problems."[15] These problems were never specified, however. More recently, the state of Massachusetts examined emissions trading in an acid deposition context as part of the federally funded STAR (*ST*ate *A*cid *R*ain) program.[16] The study, which lasted 15 months, addressed a wide variety of program and administration issues within that state. It concluded that emissions trading appears promising for acid deposition, and foresees an expanded role in the future.

One of the more ironic aspects of studying emissions trading in an acid deposition context is the fact that utilities, which have supported the bubble and other regulatory reform policies, do not include comprehensive trading provisions in their cost analyses, while the environmentalists (who are either opposed to the ETP or provide lackluster support) to minimize potential costs include full interstate trading provisions in theirs. While both sides can be expected to put their viewpoint in the best possible light, their adversarial stance also often focuses on the non-economic concerns discussed in Chapter 1. ELI notes that market-oriented systems go against the principle that pollution emitters do so at the sufferance of the state. On the other hand, a utility spokesman was quoted in the *Wall Street Journal* as saying that a few environmentalists "appear absolutely paranoid that someone will make money from cleaning up a facility."[17]

ECONOMIC CONCERNS

A wide variety of policy options is available for controlling acid deposition, and extensive economic and cost impact studies have already been undertaken.[18] Most of the cost analyses of acid deposition control options, including those involving emissions trading, have been addressed by one of two policy models—ICF's Coal and Electric Utility Model (CEUM) and the Teknekron Research, Inc., Utility Simulation Model (USM).[19] The models are both quite sophisticated, with extensive data bases and submodels to perform various coal supply, transportation, financial, and pollution control operations. An extensive evaluation of the mathematical algorithms employed in each model, and their use in setting acid deposition policy, has already been undertaken by Silverman.[20] He notes that the two models, despite seemingly sharp differences in decision rules, are quite similar in approach: both use the simulation solution to the least cost investment and operating schedule problem.

The cost minimization portion of the models is the cornerstone of this approach and has drawn attention in a number of ways. For example, attempts to use the models for analyzing the emission tax acid deposition strategy discussed earlier are not completely on target, since the regulatory constraint (that is, the tax) "effectively distorts the relative price of inputs to the firm such that the cost minimization from the firm's perspective involves an excess use of capital."[21] This distortion, based on the Averch-Johnson effect, is not believed to be particularly large, however. Of special concern to the electric utilities is the aggregated nature of the ICF model. The model minimizes costs on a nationwide basis, and any disaggregation occurs on the state level. Thus, the model can distinguish between inter- and intrastate emissions trading, but not between intrastate (external) trades and system bubbles (that is, intrafirm, internal trades).

While the model's structure and data aggregation may affect cost estimates of the various options, it is also important to keep in mind another issue that is as important as economic efficiency (and perhaps more so)—the distributional impact of the policy options. Hahn and Noll, in an evaluation of a marketable permit system for SO_2 in the LA basin, found that "the effects of a change in policy on the distribution of wealth can be very large in comparison to the improvement in economic efficiency."[22] Specifically, they found that the creation of a new property right (the marketable emission permit) would be "roughly ten times as valuable as the short-run efficiency gains."[23]

Distributional impacts are not solely a problem with economic approaches, of course. The distributional impacts of the other economic mechanism, emission taxes, are discussed in Chapter 2. The "command/control" NSPS scrubber requirement for utilities, and its "clean coal/dirty air" coalition of sponsors, has received considerable attention in the literature.[24] Any similar regulatory option (for example, mandating use of scrubbers) is likely to receive backlash because of this experience, although many of the same ends might be served by "local coal" provisions.

The distributional impact is primarily a political concern, and the debate on acid deposition controls has not surprisingly focused in that direction. For example, the Office of Technology Assessment (OTA) evaluated the impact of an emission "cap" that would place a maximum limit on emissions per unit of energy input (for example, 1.5 lbs. SO_2 per MM BTU heat input). This approach would have its greatest impact on those states emitting relatively large amounts of pollutants per unit of energy consumed, such as Missouri, Indiana, Ohio, and Tennessee.[25] Three variations of the cap were analyzed: (a) one that gives credit for reductions taken within the past 5 years; (b) one that requires a maximum reduction of 50 percent below current levels; and (c) one that requires a maximum reduction of 50 percent below SIP levels. The first and second variation would be favorable to Ohio; the third would penalize the state. The debate has not been concerned with the economic efficiency of these options, however, but rather with the distributional impacts.

ORGANIZATIONAL CONCERNS

While economic modeling is a useful approach to determine probable costs, one of the ICF analyses notes that "under each regulatory alternative, utilities are assumed to meet the given constraints (emission rollbacks, taxes and/or local coal protection) in the least-cost method possible. Difficulties with the implementation or enforcement of each alternative could limit the ability of utilities to pursue these least-cost strategies."[26] Implementation difficulties with the emissions trading program have been discussed in a previous section; cost estimates that rely on the free trading of emissions reduction could well be understated.

An even more significant policy problem can be suggested by turning ICF's quote completely around. This would imply that difficulties in the implementation of cost-effective regulatory strategies (that is, emissions trading) could be caused by the failure of utilities to pursue least-cost strategies. To put it quite simply, factors other than economics may play a more important role in utility decision-making, and this is likely to have an impact not only on the cost of any acid deposition policy options, but also on emissions trading itself. The remainder of this section discusses this particular issue, focusing particularly on organization within an electric utility.

An especially relevant guide for this type of analysis is Roberts and Bluhm's *The Choices of Power: Utilities Face the Environmental Challenge*. This book presents an in-depth examination of the way decisions concerning environmental control actions were made at six electric utilities. The authors state that it "represents an attempt to contribute to the revival of an older school of economic analysis, now largely displaced by formal mathematical theorizing based on the assumption of consistent profit maximization and on statistical investigation."[27] They go on to note that

> this older methodological tradition of non-quantitative theorizing and detailed case study evidence can contribute greatly to the understanding of social and economic processes. It emphasizes the diversity of managerial motivation, and thus facilitates the empirical study of specific, complex decisions that are inexplicable in terms of simpler models and assumptions.[28]

It thus appears to be quite appropriate for investigating organizational concerns.

The authors present a framework with four major components:

1. the external conditions that organizations face;
2. how the internal structure and operating practices transmit these pressures into decision problems facing organization members;
3. the attitudes, preferences, and beliefs that influence the way in which members solve these problems; and
4. how the structure of the organization aggregates these individual decisions into a pattern of organizational action.[29]

As discussed to a limited extent in previous sections, the utility sector is currently subject to external pressure in a number of different ways:

Uncertain Growth Projections. The early 1980s exhibited a slowdown in the growth of electrical demand, but it is not clear how much of this was due to economic recession; some believe that the slowdown was only temporary and that growth rates will increase (perhaps dramatically) in future years.

Long Lead Times for Construction of New Capacity. At the same time that projecting demand has become more uncertain, the time required to bring new capacity on line has been increasing.

Financial Pressures. Delays in construction (due to permitting and changing regulations), high interest rates, and the capital intensity of the industry have led to a precarious financial position for many utilities.

Regulatory Climate. Consumer concern over rapidly escalating rates has led some PUCs to take unfavorable positions for the utility investors. This in turn exacerbates the financial position of the firm and its future costs of raising capital.

Increasing Calls for Deregulation or Reregulation. With many individuals feeling that the existing regulatory structure is unsatisfactory, there have been attempts to make PUC Commissions subject to popular vote; to require public referenda on major utility decisions; to substitute a regional regulatory body for state PUCs; and to allow competition in electrical generation while maintaining the transmission/distribution system as a regulated monopoly.

The potential for acid deposition control legislation is also an external pressure. If an acid deposition control program were adopted, the other external pressures do not go away; in fact, they are only exacerbated. The acid deposition control program would, from the utilities' viewpoint: (a) require large amounts of "non-productive" capital; (b) result in rate increases that would further arouse consumers' ire and cause potential problems with the PUCs; (c) lead to a decrease in demand for the higher priced power; (d) potentially lead to derating of existing capacity; and (e) make cogeneration or generation from sources other than the utility more attractive.

The manner in which the utilities would shift these pressures into decision problems facing organization members (that is, the second component of Roberts and Bluhm's framework) is obviously a function of the individual utility. When faced with the large environmental control costs of the early 1970s, utilities responded in a variety of different ways. For example, TVA "strenuously refused to meet state regulations limiting sulfur dioxide emissions, and in the past sought to minimize its environmental expenditures on many occasions."[30] The study found that "politically and legally TVA has been relatively insulated from external pressures [and] has often been able to make its own judgments regarding the wisdom of environmental expenditures."

On the other hand, they note that Pacific Gas & Electric has "acquired a national reputation as an industry leader in environmental consciousness. It has developed the diverse perspectives internally that allow it to propose innovative responses to external demands."[31] PG&E has responded to the external pressure with what Roberts and Bluhm call "positive responsiveness." The firm has been willing to experiment with emissions trading in the existing ETP (as described earlier), and would probably do so under additional emissions constraints.

These examples indicate the range of responses that could be expected under an acid deposition control program—from continued resistance through progressive compliance. Some utilities might even play a role in developing the policy. Roberts and Bluhm note that the organization often gains control over the timing, the extent, and the nature of the requirement by becoming involved early, and can reduce future uncertainty, if nothing else.[32]

Jaedicke has already noted "a marked increase in strategic planning activity among utility companies."[33] To the extent that more firms are conducting strategic planning and are willing to examine cost-efficient options, the use of emissions trading would be enhanced. As noted below, however, cost is only one factor in the strategic picture, and a number of firms that have been recalcitrant in the past would probably continue so under an acid deposition program as well.

Several other factors are important when considering the manner in which utilities shift the external pressures into decision problems facing organization members in an acid deposition context. The probable costs of the control program are sufficiently high that they will receive attention from the highest management ranks. With this advantage, the issue will not be viewed as strictly a technical or economic problem, but rather as a major strategic question for the firm.

The management's options will still be influenced by the detailed technical, economic, and legal considerations determined by individual specialists within the firm, however. Technical aspects of SO_2 and NO_x control options have been discussed extensively in the literature.[34] Each option must be considered for every individual plant site and unit, and factors such as mine-mouth plant location and retrofit factors (that is, space limitations) will be critical. The options not only present different costs to management, they also present different elements of risk. The economic and legal implications can also be important. For example, fuel adjustment clauses might favor a shift in fuels rather than capital investment in scrubbers; yet the terms of a long-term contract for coal may present a major impediment to the fuel switch.

The models used for estimating costs discussed in the beginning of this section make an attempt to capture as many of these factors as possible. They put a premium factor on retrofits (for example, they assume it costs 1.3 times as much to retrofit a scrubber as it costs to install one on a new plant) and even treat long-term contracts as a model constraint. They are not able, however, to capture the third component of Roberts and Bluhm's framework: the attitudes, preferences, and beliefs that influence the way in which utility managers solve these problems.

A classic example is the problem of SO_2 scrubbers. Highton and Webb have noted that a regenerable FGD scrubber "is a complex process which effectively converts power stations into chemical processing plants for which somewhat different knowledge and expertise is required."[35] Thus, a management's decision to install this type of scrubber entails not only cost and engineering considerations, but may also affect the very nature of the manager's job. One utility executive said: "I am not in the chemical business and I don't want to be."[36]

The manager's background may also exert a certain influence on his decision. Roberts and Bluhm note that "organizations whose personnel practices tend to put lawyers in top management jobs may act quite differently from those whose recruitment and advancement systems turn key tasks over to engineers."[37] An enthusiastic endorsement of emissions trading would be expected primarily from economists. Utilities have traditionally had conservative management, with long periods of in-house grooming for management roles. While economists have made recent inroads into management, the key decisions about emissions trading are more likely to be made by lawyers and engineers.

The final component of Roberts and Bluhm's framework considers how the structure of the organization aggregates these individual decisions into a pattern of organizational action. In their analyses of the utilities, they found that "In the real world . . . problems are far too subtle and complex to be solved by formal mathematical analysis. Instead, the companies approached them by relying on habit, in what can perhaps be best characterized as a 'strategy-governed' manner."[38] Thus cost (or cost effectiveness of an action) is only a portion of the decision-making process. They note: "Not every organization was or perceived itself to be tightly constrained by cost . . . Cost-consciousness was thus in part a matter of regulatory constraint and the larger strategic picture. What appeared to be cheap or expensive to each organization depended heavily on their expectations, which changed over time."[39] This finding is consistent with the observations about utility participation in the ETP discussed in Chapter 3. The potential for future regulatory actions has a much stronger bearing on a utility's actions than any near term efficiency gain. It is also consistent with the historical emphasis on system reliability.

The fact that utilities are not necessarily bound by cost minimization has implications not only for the economic modeling analyses discussed earlier, but also for the emissions trading program (since that is its prime reason for existence). It implies that efforts to enhance use of the market should focus on minimizing the strategic uncertainty in the program.

THE UTILITIES' VIEWPOINT ON EMISSIONS TRADING

To summarize briefly our findings to this point, a review of EPA's Emissions Trading Program from an electric utility's perspective indicates that there are two kinds of impediments to emissions trading: those that utilities share with industrial firms in general, and those that are utility-specific.

The shared impediments include market constraints, uncertainty, transaction costs, and other related concerns, such as the general economic climate. Utility-specific impediments include rate-making risks, longer construction planning times, the potential need for ERCs for customers, and an enhanced likelihood of future regulatory requirements. These impediments have led to the hoarding of ERCs.

The use of some emissions trading by electric utilities in an acid deposition control program seems likely, but the decision to participate in the market will be affected by numerous external pressures, organizational aspects of the individual utilities, and even the behavioral characteristics of the utility management. Moreover, one major factor is that utilities are not particularly oriented toward minimizing cost, placing greater emphasis on the strategic aspects of the control option decision. Concern for system reliability and future regulatory uncertainty are considerably more important than near-term efficiency gains.

With this viewpoint, it appears that intrafirm, inter-plant emissions trading (that is, systems bubbles) would be used by electric utilities, since this gives management additional flexibility with little added risk. Further steps toward interfirm trading on an intrastate basis would not have major political problems, but it is not clear that management would use the option extensively. Interstate trading was viewed as having both political and utility-specific problems, with the possible exception of interstate trades that are a part of a systems bubble (for example, for utilities such as TVA, American Electric Power, etc.).

It is apparent from this review that the current level of activity in the market leaves much to be desired, and emissions trading would probably not represent major cost savings in an acid deposition control program. To stimulate market activity, one must recognize both the market impediments and the desire of utilities to address the strategic implications of the control technology. One potential approach would be to increase ERC leasing.

Part II

◆

The Survey

5
Survey Design

Numerous analyses discussed earlier have estimated the cost of acid deposition controls, assuming that emissions trading would occur, despite the fact that there were no data available concerning (a) the attitudes toward emissions trading and ETP experience of the utilities who would have to do the trading, and (b) the attitudes and experience of Public Utility Commissions in the area likely to be significantly affected by acid deposition legislation. This section outlines the results of a survey designed to address these issues, and explores the relative importance of the market impediments outlined earlier. It also reveals both utility and PUC reactions toward ERC leasing, which could represent an approach for increasing market activity.

The questions in the survey included the following:

> Have you had any experience with EPA's Emission Trading Program? Were there any major impediments (either in-house, from the regulatory agency, or from outside groups)?
>
> Some electric utilities may have been reluctant to sell "Emission Reduction Credits" (ERCs) because of uncertainty about future control requirements. One proposed solution has been a reliance on more varied leasing arrangements for ERCs, so the utility can maintain control over future use of the ERC. Do you think leasing ERCs would improve the likelihood of emissions trading?
>
> Some studies have assumed that full emissions trading across a 31-state region is feasible; others say that interstate trading is "totally inconsistent with political reality." Do you believe that interstate emission trades between electric utilities are feasible?

Personal visits or telephone interviews to discuss these questions were conducted with 23 utilities and 14 PUCs. This chapter describes the survey design. Chapters 6 through 8 will describe the results for each of these questions in turn, first outlining the responses of the utilities and then those of the PUCs. Chapter 9 will discuss the implications of the survey results on market activity.

One must keep in mind while reading this section that those interviewed constitute two relatively narrow groups. The viewpoints of these specific groups were solicited, and the survey was intended to be tightly focused. Certainly, environmentalists and state air pollution agency representatives would give considerably different responses. The viewpoints outlined in this report are those of the survey participants and are not necessarily those of the authors.

The selection of the utilities and PUCs for the survey was influenced by several factors. First, more utilities were included than PUCs. Utilities are considerably greater in number, were perceived as being more familiar with emissions trading, and would be the organization actually initiating an emissions trade. Second, emphasis was placed on utilities with coal-fired capacity. Third, utilities of all sizes were contacted. Fourth, other characteristics (for example, municipal ownership, interstate generation, etc.,) were considered to ensure that responses would address these factors as well. It was also important that the survey focus on that geographic area most heavily affected by proposed legislation. Table 5.1 gives a ranking of the six states with the greatest coal-fired capacity, total SO_2 emissions, and potential emissions reductions required under Senate Bill S-3041. It is important to note that four states—Ohio, Pennsylvania, Indiana and Illinois—are included on all three lists; those states therefore formed the core of the survey.

A list of utilities in these and nearby states was developed, based on coal-fired capacity. The resulting capacity in each state is listed in

Table 5.1 - State Rankings

	Existing coal-fired capacity[a]	Total SO_2 emissions[b]	Potential reductions under S-3041[c]
1	Ohio	Ohio	Ohio
2	Pennsylvania	Pennsylvania	Pennsylvania
3	Illinois	Indiana	Missouri
4	Indiana	Illinois	Pennsylvania
5	W. Virginia	Missouri	W. Virginia
6	Kentucky	Wisconsin	Illinois

[a] Based on 1981 data in U.S. DOE, Inventory of Power Plants in the United States: 1981 Annual. Energy Information Administration, DOE/EIA-0095(81), Sept. 1982.

[b] Based on 1980 data in OTA, The Regional Implications.

[c] Based on 1995 estimates, assuming intrastate trading, in ICF, Inc., "Analysis of a Senate Emission Reduction Bill."

Table 5.2. This shows that approximately 40 percent of coal-fired capacity in the 31-state region is represented in the survey. For the four core states, 80 percent of coal-fired capacity or greater is represented. Nearby states that were included in certain listings in Table 5.1 are also well represented.[1] The size distribution of the utilities in terms of coal-

Table 5.2 - Survey Participants

State	Existing coal-fired capacity (MW)	Existing coal-fired capacity of utilities in survey (MW)	Percent coal-fired capacity	PUC contacted
Alabama	10585	-	0	
Arkansas	2112	-	0	
Connecticut	-	-	-	X
Delaware	743	-	0	X
District of Columbia	-	-	-	
Florida	4857	-	0	
Georgia	11479	-	0	
Illinois	17314	15421	89	X
Indiana	16164	12922	80	X
Iowa	5614	-	0	
Kentucky	14434	1765	12	X
Louisiana	1118	-	0	
Maine	-	-	-	X
Maryland	3034	110	4	
Massachusetts	-	-	-	X
Michigan	11329	-	0	
Minnesota	5076	-	0	
Mississippi	2428	-	0	
Missouri	12088	-	0	X
New Hampshire	459	459	100	X
New Jersey	1665	299	18	X
New York	2977	-	0	X
N. Carolina	11821	-	0	
Ohio	24031	19734	82	X
Pennsylvania	19291	19158	99	X
Rhode Island	-	-	-	
S. Carolina	4126	-	0	
Tennessee	10021	-	0	
Vermont	30	-	0	X
Virginia	3488	-	0	
W. Virginia	14779	12051	82	X
Wisconsin	7233	5203	72	X
Total	218,296	87,131	40%	

Table 5.3 - Size Distribution of Utilities in Survey

MW coal-fired capacity	Number of utilities
0 - 500	3
500 - 1000	2
1000 - 2000	2
2000 - 3000	5
3000 - 4000	5
4000 - 5000	-
5000 - 7500	4
7500 - 10,000	1
> 10,000	1

fired capacity is outlined in Table 5.3. The capacities of all subsidiaries are included with the parent firm.

The PUC at each of the eight states listed in Table 5.1 was contacted in the survey, as well as other states (primarily in the Northeast). PUCs included in the survey are also outlined in Table 5.2. The coal-fired capacity in these states represents approximately 60 percent of the total coal-fired capacity in the 31-state region. The majority of contacts at electric utilities were provided by the Edison Electric Institute. At the PUCs, it was necessary to call "blind" and attempt to locate the person most appropriate to discuss the survey questions. When the name of an appropriate individual was obtained, a letter outlining the survey questions was sent. Four categories of letters were mailed: (a) eight to electric utilities requesting a personal interview; (b) sixteen to electric utilities requesting a telephone interview; (c) five to PUCs requesting a personal interview; and (d) twelve to PUCs requesting a telephone interview. Edison Electric Institute had predicted negative responses to the requests for interviews, but the results were extremely positive. In the electric utilities, all eight persons contacted agreed to personal interviews, and fifteen of the sixteen agreed to be interviewed by telephone. The results of the PUC contacts were nearly as successful. Four of the five persons agreed to a personal interview, and eleven of the twelve PUCs were contacted by telephone. One of the PUC representatives, who gave monosyllabic replies to the four questions, was considered nonresponsive.

Interviews were thus obtained in 37 of the 41 attempts (23 of 24 utilities and 14 of 17 PUCs). While the letter requested no more than one hour for a personal interview and twenty minutes for a telephone interview, many personal interviews lasted two hours or more, and several phone conversations lasted more than one hour. Since confidentiality was promised, the utilities and PUCs are not mentioned by name in this study.

6

Experience with Emissions Trading

The only major compendium of transactions in EPA's Emissions Trading Program was the analysis performed by Vivian and Hall at the University of Michigan's Institute of Public Policy Studies.[1] The first question in our survey updated this information in the study area and was a measure of the respondents' familiarity with the emissions trading concept.

ELECTRIC UTILITIES

When asked whether the firm had any experience with EPA's Emissions Trading Program, only one electric utility responded affirmatively. It had banked emissions from the shutdown of an oil-fired boiler used for steam heating in an urban location; the operation had lost money as individual buildings in the area switched to natural gas. The area was one of several in the country where a local agency had developed a formal emissions bank, and the ERC had been banked for approximately one year. This constituted the only formal use of an emissions bank found in the survey.

Nonetheless, the lone initial positive answer masked a wide range of activity, which became apparent after further discussion. The term "EPA's Emissions Trading Program" connoted to most utility representatives the use of offsets and emissions banking; they viewed bubbles more as a flexible permitting mechanism. Thus, one utility representative stated in the beginning of his interview that his firm had no experience with the ETP, but he later described three bubble applications the firm had submitted to the state air pollution control agency.

While the full connotation of the term "Emissions Trading Program" was not apparent to all, every utility representative interviewed was familiar with the individual components of emissions trading. The use of terms such as "banking," "offset," "bubble" and "ERC" never elicited questions. This is not surprising, since the primary contact in the firm was usually in the environmental affairs department. In the course of the survey, however, we spoke with lawyers, operating engineers, meteorologists, and analysts in systems planning departments, and all were conversant with the topic. This gives some indication of the extent to which emissions trading has already become integrated into utility operations.

This chapter describes, from several viewpoints, the experiences of utilities with emission trading, including internal analysis, growth considerations, contacts with outsiders requesting ERCs, etc. It then discusses the two principle types of ERCs—those associated with the overcontrol of emissions, and with unit retirements. The persons interviewed were not reluctant to discuss their opinions about problems with the ETP and the potential applicability of trading in an acid deposition control program, and these viewpoints are documented as well.

Case Histories

Several utilities had made proposals (with various degrees of formality) to air pollution agencies concerning trading options. For example, one midwestern utility had submitted three bubble applications to the state's air pollution agency. All three applications concerned particulate emissions. In two of the applications the utility had high-efficiency electrostatic precipitators (ESPs) on certain units and wanted to modify these control devices to decrease emissions, with a corresponding increase in allowed emissions from other, older, controlled units. The third application was for an external bubble. This plant was in an economically depressed area, and the utility had talked to a number of other industrial firms about the availability of particulate ERCs from shutdown facilities. It is interesting that one industrial firm offered to lease the ERCs, but the utility was able to locate other available ERCs within five miles of the plant site. It had applied to the state to use the ERCs to increase emissions from its existing plant. Preliminary prices for the ERCs had been discussed, but the exact price would not be determined until their applicability had been defined. At the time of the interview the state had not yet acted on the applications.

Another utility had discussed two potential bubbles at its generating plants with the state air pollution control agency; one plant had two units that had to meet an identical SO_2 emission limitation. With their existing operations, one unit was consistently under the required level, while the other was consistently over. Thus, the utility would have preferred to treat the two units as one bubble. The state, however, had taken the position that the utility should "clean up its act" for the noncompliance unit through better fuels management and coal-handling equipment. The utility spokesman showed some frustration with the state's position: "even though we are offering over-control which is both permanent and quantifiable, the state people continue to refer to these proposals as 'relaxations.' " The second discussion on a potential bubble focused on particulate emissions from a different generating station within the system. The firm had three ESPs, one emitting slightly more than permitted, a second on target, and a third below allowable levels. The state indicated that it did not want to consider the proposal until applicable bubble regulations had been finalized.

As noted earlier, one utility had banked emissions from a shutdown steam heating facility. A group of local building owners planned to convert the oil-fired unit to natural gas, and the utility was discussing the potential release or sale of the banked ERCs (or part of them) to the group so they could net out of PSD review.

Finally, two other utilities interviewed had submitted proposals that would expand the boundaries of the ETP somewhat. The existing bubble regulations are applicable only between existing sources. New sources are required to meet emission limitations specified by New Source Performance Standards. In the past, an attempt was made by Asarco to use the bubble between an existing source and a new source, but this was struck down in court.[2] Recently, however, the U.S. EPA has discussed allowing a new source/new source bubble by the Central Illinois Public Service Company (CIPS). CIPS has two 590-MW coal-fired units affected by the original 1.2 lbs. SO_2 per MM BTU limit set by the NSPS. The proposed bubble would allow it to operate one unit at 1.8 lbs. SO_2 per MM BTU and the other unit at 0.6 lbs. SO_2 per MM BTU to meet the 1.2 requirement. The utility estimated that the proposed bubble could save its customers $10 to $15 million every year for the 30-year life of the facility.[3]

The Pennsylvania Electric Company (Penelec) has gone a step further and proposed that a new source/existing source bubble be allowed at its Homer City plant.[4] That plant has three units, two limited to 4.0 lbs. SO_2 per MM BTU by the SIP, and one to 1.2 lbs. SO_2 per MM BTU by the NSPS. The proposed bubble (termed a "continuous emission offset" by the utility) would include all three units; the plant was currently operating under such an arrangement as part of a consent decree with the Pennsylvania Department of Environmental Resources. While the Asarco new source/existing source bubble was struck down by the courts, Penelec's proposal took into account the major problem. Penelec's bubble would change over time to account for unit shutdowns and/or retirements. Thus, when the two existing units would be retired, the required emission limitation for the NSPS unit would revert to 1.2 lbs. SO_2 per MM BTU, and the new unit would not be allowed to emit at levels above those equivalent to previous control requirements. The utility estimated that resultant savings through use of the bubble would be greater than $2 million per month.

Thus, utilities in the survey had banked emissions, submitted bubble applications, held discussions with state agencies concerning proposed emission trading options and, in two cases, submitted proposals that would extend the boundaries of the ETP.

Nor had the remaining utilities ignored the ETP. Since familiarity with ETP options was widespread, most of those interviewed had given at least cursory thought to applying the options to their own system. Only two firms indicated that they had not considered any options, and both gave the same reason: their plants were already in compliance with existing regulations. A number stated that evaluations were performed strictly on an internal basis, and many said that these evaluations were very preliminary, "back-of-the-envelope" kinds of calculations. One firm indicated that it was in the process of employing an outside consultant to review emission trading options, as well as other potential courses of action (for example, petitioning for a SIP change and doing dispersion modeling studies).

Firms in at least two different states indicated that they had had experience with the bubble, because it was not new; the state SIPs had included rules similar to the bubble provisions before the policy was adopted. One example was the compliance equations promulgated by the U.S. EPA in the Ohio SIP in 1977–78.[5] These equations give the utility some flexibility in meeting the emission limitations, and affected virtually all power plants in that state.

Several firms indicated that they had considered emissions trading in terms of future growth. One discussed the problem of siting new plants and indicated that his firm had evaluated sites in terms of the need for offsets (that is, the site's proximity to nonattainment areas). He ventured the opinion that most utilities would rather deal with PSD regulations than offsets, and would specifically go out of their way to avoid offsets if possible. Another firm stated that it would use ERCs to whatever extent possible to attract new industry into the area, and had already had discussions with a major industrial firm concerning needed offsets.

Indeed, one of the interesting results of the interviews was the response of utilities who had already had discussions with outsiders concerning ERCs. Two of the cases have been described earlier (the steam-heating case and the external bubble application). Four other utilities had been approached by outside sources. In one case (mentioned immediately above), the firm indicated that it would use ERCs in whatever manner possible to foster industrial growth in its service area. In a specific case, the utility met with an industry proposing to build a new plant in order to determine what ERCs would be required.

It ultimately became clear that no ERCs were necessary, but the utility had begun to evaluate options in any event.

The three remaining cases were considerably different. In one case, a utility with a generating station in an urban area was approached by an industrial firm looking for SO_2 ERCs for an offset. The industrial firm apparently had already approached a number of other SO_2 sources in the area, but had not been able to locate anyone willing to sell ERCs. What makes the case remarkable is that a number of the firms approached had plants that had been shut down for at least a year. The sale of ERCs would have interfered with the utility's plans for its urban station, but the spokesman indicated that they would not have sold ERCs even if they had them. The existing firms, realizing that they controlled the future economic development of the area, were unwilling to take a chance on losing something that might become much more valuable later. The utility recognized this—its own thoughts were similar—but was nevertheless concerned about the impact this was already beginning to have on industrial development in the area. The utility could thus see itself in a Catch-22 situation—hoarding ERCs and hindering economic development, or selling them and losing control over its own future actions dictated by the limited air resource. The spokesman noted that the utility would also protect its future options by holding the ERC in this type of case.

A second utility was approached when it was preparing to install a new ESP, and was asked about the feasibility of providing particulate ERCs through overcontrol. The proposal "never got beyond the talking stage." The utility indicated that there were a number of legal and contractual impediments; that it was very difficult to determine the responsibility of each party; and while emissions trading was good on a theoretical basis, it was not practical. The firm indicated that it was not interested in selling ERCs in any event.

A third utility had installed a flue gas desulfurization (FGD) system that provided some control beyond allowable levels, and the firm had been approached by an industrial firm that wanted to "borrow" the potential ERC. The utility had indicated that it would not sell the potential ERC from its overcontrolled emissions.

Thus, of the six potential external ERC cases in the survey, in one case the utility had approached outside sources; in the other five, the utility had been approached. In three of these five cases, the utility indicated that it would not make ERCs available for sale.

The results of the utility experience with emissions trading are summarized in Table 6.1. There is little formal use of the ETP. The utilities were thoroughly familiar with emissions trading regulations, however, and had integrated them into their operations. The survey data confirmed the internal bias of the market, that (a) most activity

Table 6.1 - Survey of Electric Utility Activity in the ETP

	Number of Utilities
Formal use of ETP:	
Banking	1
Offsets	0
Bubbles	1[a]
Analysis of ETP provisions for internal applicability	21
Discussions with state agencies concerning potential use of ETP	3
Potential external transactions:	
Approached by other firms	5
Approached other firms	1
Total number of utilities interviewed	23

[a]Application for 3 bubbles.

was focused on internal applicability, and (b) the response was predominantly negative when the utility was approached by outside firms concerning ERC availability. Only one firm had made an effort to seek external ERCs. Most of the external activity had focused on plants located within or reasonably close to a metropolitan area, where the incidence of emissions trading was likely to be higher than average. Given the high visibility of utilities as an emission source, it seems probable that they would receive inquiries often if ERCs were needed in that area. The fact that only 22 percent were contacted by outside firms is perhaps indicative of the economic problems in the late 1970s and early 1980s in the Northeast and Midwest.

Overcontrol

Two of the utilities discussed above had the potential for ERCs based on overcontrol, and a number of other utilities indicated that they also had this capability. In most cases the discussion on overcontrol focused on particulate emissions, although two utilities indicated that they were able to overcontrol on FGD systems. Most indicated that they viewed overcontrol as a "cushion" and were not interested in pursuing its applicability in the ETP. One utility spokesman noted: "We have a cushion in our operation, but I wouldn't want to bank on it . . . no pun intended."

The utilities felt that having the cushion was important for negotiations with state air pollution agencies. One noted, "I could lose that margin with the stroke of a pen. All it would take is a different averaging period for the sulfur content of the coal." The desire to maintain a cushion would have important implications for emissions trading in an acid deposition context, since overcontrol would be an important source of ERCs.

Previous experience with overcontrol was not always positive. One utility had installed electrostatic precipitators at a plant site as early as the 1930s; when the original SIP was developed after the 1970 Clean Air Act, the utility overcontrolled with ESPs while meeting opacity requirements. The area was designated as nonattainment for TSP after the 1977 Clean Air Act Amendments, and the overcontrol was taken (that is, made a required emission limitation). One problem was that the overcontrol was based on high-sulfur coal. When ambient SO_2 levels were identified as a problem, the utility was required to shift to low-sulfur coal; the resistivity of the low-sulfur coal caused a decrease in ESP efficiency, and the particulate emissions exceeded the allowable levels. The utility was therefore required to spend millions of dollars on an ESP upgrading program. The spokesman noted: "In retrospect we made a mistake in overcontrolling—we won't make that mistake again."

Others who have attempted to use the ETP for overcontrolled emissions have not always been successful. One utility had overcontrolled particulates at a remotely sited plant. The firm was concerned about future expansion of the plant site, and since the PSD baseline would be set on actual—not allowable—emissions in the area, the firm decided that it wanted to bank the overcontrol. Nevertheless, the state agency said that the area was remote and was attaining the NAAQS, and that there was no demand for ERCs in this type of area; hence, no bank was required. The firm made a distinction between the only PSD offset case in the country to date (one in North Dakota, where the entire PSD increment was consumed) and their case, which merely affected the determination of the baseline. The firm had documented the overcontrol, but was uncertain of its future applicability in regulatory actions. Overcontrol appears to be a critical component of emissions trading in an acid deposition program, and has been shown to be practical from the number of cases documented. It has not made significant inroads in the ETP, however, because of the utilities' desire to maintain a cushion, a perception that the overcontrol could be easily lost or made mandatory, and the apparent reluctance of states to give ERCs for the overcontrol.

Retirement

Several firms indicated that their principal interest in the ETP was the potential for banking emissions from retired units. As noted earlier, the only formally banked ERC documented in the survey was from a shutdown unit. A number of firms were uncertain about the applicability of these ERCs, since some states included plant shutdowns in their SIP attainment strategies, and others were only allowing ERCs from shutdowns to be used in conjunction with new source growth or plant expansions.

One firm indicated that it had given some thought to obtaining ERCs from unit shutdowns, but decided instead that it was in its best interest to keep the permits and obtain the ERCs at a later date when they were needed for future control requirements. The firm had considered banking and had contacted the state's air pollution agency, but was told that ERCs had to be used within five years or would be lost. The firm indicated that it was not operating the retired units; when asked whether it was losing its ERC in a de facto manner by having zero emissions in recent years averaged into the ERC determination, the utility responded that it did not think so but was uncertain. Subsequent discussions indicated that zero emissions were not being added, but the five-year clock was running. Thus the utility was effectively banking its emissions without using the state's bank. The extent to which other utilities are employing this technique, and thus creating an understatement of utility activity shown in Table 6.1, is not clear. Most firms indicated that they had not included premature unit retirements in any acid deposition control analyses; one said that this would occur only if the control requirements were extremely onerous.

Impediments in the ETP

Many of those interviewed provided extensive comments about impediments in the existing ETP. Some comments were more general than the items previously identified: "Emissions trading is great in theory, but it's just not practical." "Emissions trading might work about 15 years from now." Other concerns were more specific, raising issues such as how to deal with scrubber variability, outages of a unit within a bubble, appropriate averaging times for the ERC, and the determination of actual versus allowable emission rates in the SIP baseline. Most comments could be placed in the impediment categories described earlier, however, and are therefore grouped below in the same manner.

REGULATORY CONSTRAINTS

The overwhelming number of comments in this area concerned the utility's relationship with the state air pollution control agency. Problems concerning the use of the term "relaxation" and difficulties in banking ERCs in an isolated attainment area have already been described. As noted earlier, different states have different philosophies about emissions trading, and utilities in at least two states indicated that, in their opinion, the state agency was not particularly in favor of the ETP. This is certainly not a one-way issue. One state agency was termed as "progressive," but a utility within the state indicated that it was testifying against adoption of an alternate emission reduction (bubble) rule because there were not yet sufficient cases to show whether the rule would work favorably; they did not want the state to proceed until more detailed requirements were issued by the EPA headquarters in Washington.

Another utility spokesman argued with EPA's contention that the ERC was a currency in the market; he held that the characteristics of each ERC and its constrained use were so specific (that is, non-fungible) that the ERC was not a currency, but rather the commodity itself.

UNCERTAINTY

Several utilities mentioned that their states did not yet have formal banks in place, and one stated that "right now there is just no legal protection of an ERC asset." As noted earlier, another stated that legal and contractual requirements were a problem, and that it was difficult to assign responsibilities to each party.

The "moving target" of regulations was similarly commented on by a number of utilities; one admitted that they view all their emission limitations as "a point for present discussions only—the regulations have changed in the past and they're going to change again in the future." One spokesman was emphatic that uncertainty was *the* major problem in the ETP. During the course of the interview he continually returned to the same theme: "the market has no stability . . . Things are always changing, and they're subject to political whim . . . The whole program is too tenuous, and ERCs could disappear too easily."

TRANSACTION COSTS AND OTHER IMPEDIMENTS

Only one person interviewed mentioned transaction costs, and gave the specific example of the Sohio/Southern California Edison case. Economic conditions, and the declining need for new baseload units—often until after the year 2000—were cited by a number of survey participants as a reason for the slow development of utility participation in the market.

Utility-Specific Impediments

Most utilities cited the general impediments discussed above as being more important than utility-specific concerns in the existing ETP, with one exception: the enhanced potential for future regulatory requirements. As noted earlier, many felt that upcoming acid deposition control requirements would have an impact on their actions. For example, one of the utilities that had overcontrol on an FGD unit noted that they would not sell the potential ERC at all, since they would probably need it later. Similarly, the firm that held permits rather than banking ERCs from retired units had potential control requirements in mind.

Rate-making risks were simply not a major issue in the existing program, although several firms thought that this could change in an acid deposition control program. The potential use of ERCs for customers was discussed earlier. The one case where the issue was raised suggests the possibility that this could be a stimulant to the market rather than an impediment, since the utility was actively seeking to find ERCs for the potential customer. The remaining utility-specific impediment—long planning times—was never mentioned by survey participants.

Emissions Trading in an Acid Deposition Control Program

When the utilities were asked whether the emissions trading approach was relevant for an acid deposition control program, their response was mixed. The positive responses were usually qualified, such as: "I think it has potential, but it will be complex" and "Yes, it's workable, but it will still be difficult when the political factors get involved." The negative responses were usually more blunt: "Trading won't work." "No, the agencies will go with what has worked for them in the past—and that means scrubbers and them setting the controls." After further discussion of the issue, however, it became apparent that these responses were focused on external transactions. The response concerning internal transactions was overwhelmingly in one direction—toward approval of emissions trading through systems bubbles.

Several firms had already evaluated the impact of using a systems bubble on their operations, and had indicated that they would use it if possible. One firm described a large bubble over a major portion of its service area, with control being focused on two specific plants. Another described its control options at a major mine-mouth plant in terms of whether a systems bubble was allowed or not.

Nonetheless, the utilities were not convinced that systems bubbles would be allowed and had not included them in most cost analyses. One stated, "Just because it seems to make sense from our point of view doesn't mean that it will be allowed." Another expressed concern about obtaining approval for bubbles between plants that were located in different airsheds, given the state's reluctance to adopt all allowable provisions in the ETP, even if the bubbles were allowed by the federal government.

Not one utility argued against emissions trading through internal bubbles between plants in the same system. Indeed, the response went beyond approval of this type of emissions trading; there was actually some fear that systems bubbles *would not* be allowed. Two utilities felt that certain plants in their system would be targeted for retrofit controls. One stated that informal discussions with EPA personnel had indicated that the utility would be required to control one particular plant (emitting well above 6.0 lbs. SO_2 per MM BTU heat input), despite the fact that the plant might avoid control altogether if the emission reduction requirement were slightly less than currently proposed levels and systems bubbles were allowed. Another utility felt that one of its plants would receive primary attention from the state agency, and it would be required to control regardless of other trading options.

Beyond targeted retrofits, one utility representative believed that EPA would not allow much latitude in the control decision. The quote above, which states that the agencies will go with what has worked for them in the past, was an initial response concerning the use of emissions trading in an acid deposition control program. After this comment, EPA's attempts to increase economic efficiency through market mechanisms were pointed out, but the utility spokesman remained unconvinced: "Well, they didn't seem too concerned with the allocation of resources when they tried to force scrubbers on us in the 1970s." Still another utility expressed concern that the final legislation might require a mandated percentage reduction for all plants, rather than allowing systems bubbles. Finally, one indicated that acid deposition controls could actually threaten the financial integrity of certain utilities if (a) the reduction allocation were made improperly or (b) if systems bubbles were not allowed.

This enthusiasm for systems bubbles did not extend toward external trading, however, and the support it did receive was qualified by concern about many of the impediments identified above, including the potential for PUC intervention.

A number of firms indicated that the PUC would not play a particularly important role, or would intervene only in certain cases. One utility representative noted: "The PUC won't play a major role in emissions trades unless, say, we were using low-sulfur coal when we could be burning (in-state) coal just to provide ERCs."

A number of respondents expressed interest in the possibility of using industrial ERCs to meet acid deposition control requirements. But the most exasperated viewpoint concerning external trades was expressed by a midwestern utility spokesman who noted that any market approach assumed that the participants acted rationally. He suggested that the "draconian measures" required in several midwestern states by proposed acid deposition control programs would threaten the very economic survival of the utilities, and the utilities would be acting under extreme financial duress. He believed that the economic theory of the market (and control cost estimates based on emissions trading assumptions) was therefore irrelevant.

PUBLIC UTILITY COMMISSIONS

While the terms and concepts of emissions trading were familiar to every utility, the same could not be said for PUCs. Initial contacts with PUC representatives showed the following results: Never heard of emissions trading, 4; Heard of emissions trading but not familiar, 6; Familiar with emissions trading, 4. The two other PUCs contacted but considered non-responsive had both heard of emissions trading; their degree of familiarity was not ascertained.

Two of the six states ranked in Table 5.1 that could require the greatest emissions reduction fall within the category of those who had never heard of emissions trading. One had circulated the questionnaire within the PUC, and no one was familiar with the topic. Both expressed interest in the program (as did the other PUCs who had not heard of the program), particularly concerning its effects on minimizing costs of an acid deposition control program.

Six of the representatives had heard of emissions trading, or had heard terms like "bubble" or "offset," usually through contacts with the air pollution control agencies in their state. Although a number were familiar with the general concepts of emissions trading, they were not familiar with the existing program. Since more than 70 percent of the PUC representatives were not familiar with the details of emissions trading, the interviews by necessity focused on some of the broader issues. Thus there were no comments about impediments in the existing ETP, but substantial comments about the mandate of the PUC in environmental matters.

Indeed, the first comment from virtually every PUC contact was that the questions should be addressed to the state's air pollution control agency. A number sent the letter directly there, and two contacted the air pollution agency before responding to the questions. One PUC conducted its own poll of several electric utilities in the state concerning the survey questions, including some utilities not covered in our listing.

Four of the PUCs were familiar with emissions trading, but only one had any experience, and then in only one case. This PUC was involved in the negotiations between a utility and the state air pollution control agency concerning a new unit. In a netting arrangement, the utility retired two units and lowered the sulfur content of fuel at two existing units to offset new emissions. The PUC took an active role in the negotiations, even requiring an increase in the offset ratio. Needless to say, the air pollution control agency was unhappy that the PUC was interfering at this level; the PUC's authority was subsequently limited by statute to avoid interference with the permit authority of the air pollution control agency.

Surprisingly, three of the PUCs had been involved in cases that would expand the boundaries of emissions trading. The Pennsylvania Public Utility Commission had followed Penelec's attempts to implement a new source/existing source bubble at its Homer City generating station, and had supported these efforts.

Utilities in another state had discussed with the PUC the potential for a new source/existing source emissions bubble at two different oil-fired plants. EPA was approached but was "neutral to negative," and apparently never acted on the matter.

Dr. Sury Putta of New York's Department of Public Service discussed this issue in a technical paper that concludes:

> The present bubble policy by EPA is of little value to the utility industry in New York and many other states where the SIPs include alternative means which allow relief from the emission standards without any emission offsets. Extension of the current bubble policy to new power plants subject to NSPS (NSPS-bubble) and to all sources within an airshed region (Regional-bubble) offer a range of policy options which in some cases result in cost savings and environmental benefits. The Regional-bubble option would help the conversion of several power plants from oil to coal burning and could be adopted by EPA under the current provisions of the Clean Air Act.[6]

Regional bubbles would extend emissions trading beyond that allowed in the ETP. The Wisconsin Public Service Commission examined a scheme that would extend the boundaries and allow for systems bubbles in an acid deposition context. The state considered potential legislation placing a cap on SO_2 emissions. It is interesting that the state found a unit-by-unit emissions cap could lead to greater emissions than a systems bubble because of the dispatching sequence.[7]

Thus, while PUC experience in the emissions trading program had been negligible, a number of PUCs were involved in issues at the policy frontier. Most were not, however. Several representatives noted that they were concerned only with the financial aspects of utility regulation, and had no mandate for environmental concerns.

When questioned about their possible participation in emissions trading and their activities in environmental issues, the survey group split evenly: seven said this would be included in their mandate, and seven said it would not. Those responding positively indicated that they expected to be very much involved in the proposed emission trade:

> If an ERC were generated by a scrubber, it could potentially lead to increased outages, changes in the heat rate and other things . . . We worry about these things as well . . . And if one of our utilities is providing an ERC for an out of state utility, we'd want to make sure that *all* of the costs are covered.
>
> We examine the variation in performance due to air pollution equipment and requirements [he then cited examples concerning outages due to opacity and derating due to scrubbers]. We'd investigate all of the control alternatives that the utility has.

Several went beyond this technical concern to discuss the environmental impacts as well: "Any action the Commission takes is evaluated for its environmental implications." "The [PUC] in this state is progressive environmentally." An equal number presented a different viewpoint: "We'd look at the control alternatives, but virtually all air decisions are made by [the state's air pollution control agency]. The PUC has no authority for environmental issues." "We deal strictly with financial and economic issues. The air pollution control agency would handle the others." "The PUC is a reactive concern. We don't go around looking for new issues." Several PUCs mentioned that any emissions trading cases would have to be handled on a case-by-case basis. One noted that emissions trading is "too new to get much feel for how it would be handled." The principal concern was that any action taken be reasonable and adequately supported for review.

Finally, a number of other miscellaneous topics were discussed by PUC representatives interviewed. One person noted that something new like an ERC might create a problem if it were not handled in a test year. (Upon reflection, however, he added that "utility rate cases are coming so often that this probably isn't a problem now.") Another noted that PUCs have not had to deal extensively with the intangible asset issue. Assets were required to be "used and useful," and he suggested that an ERC would come under an O & M budget rather than being considered an asset. Finally, one PUC spokesman expressed sympathy for the utilities. Stating that regulatory programs were very unstable, he could "understand their desire to hoard ERCs."

7

ERC Leasing by Electric Utilities

When asked about the possibility of leasing ERCs instead of buying and selling them, utilities gave a mixed response. Reactions ranged from positive: ("It's a wonderful idea . . . Something like this will have to be done in the future") to negative ("Forget it"), with intermediate reactions ("It's a good abstract idea, but not necessarily practical"). The consensus indicated that ERC leasing was an option that could be useful in certain applications, but that the number of opportunities would probably be relatively limited. Still, firms indicated that they would be willing to examine ERC leasing as a potential control option. The following sections discuss the applicability of ERC leasing by the utilities, their experience with other leasing arrangements, and the anticipated problems in an ERC leasing program.

APPLICABILITY OF ERC LEASING

ERC leasing has already been used successfully in the ETP. In Louisville, for example, General Electric leased VOC ERCs from International Harvester for a two-year period. While the use of leased ERCs in this case is a relevant precedent (as described in the next section), utility representatives raised a number of concerns about the potential supply of, as well as the demand for, leased ERCs.

Supply could theoretically be obtained in a number of ways, such as overcontrolling with a retrofit FGD system or by retiring existing units. It may also be possible for those utilities that met the original 1.2 lbs. SO_2 per MM BTU heat input NSPS with low-sulfur coal to install a scrubber to generate ERCs. Similarly, demand would exist for an older plant with a relatively low-capacity factor that could not economically support a major capital investment, and that would have difficulty in shifting to low-sulfur coal for technical or other reasons (for example, coal contracts).

While these cases appear to be relatively specialized, it is encouraging to note that during the survey such cases were found, even without any acid deposition requirement. In one case, a utility had installed an FGD system to meet a state SIP requirement that could not be met with low-sulfur coal, and the design resulted in overcontrol. The utility was approached by another source who wanted to "borrow" the potential ERC for a period of time—an ideal candidate for a lease. On the demand side, one utility with two relatively old, small coal-fired peaking units (and constrained by an existing coal contract) was currently in the process of examining emissions trading options at the plant site. In yet a third case, a utility approached a number of industrial firms seeking ERCs; one of the firms offered to lease the ERCs, but the utility was subsequently able to arrange for the purchase of other ERCs. These examples indicate that ERC leasing arrangements appear to be feasible.

UTILITY LEASING EXPERIENCE

Although utilities had not had direct experience with ERC leasing, several had leased other items in the past and had therefore dealt with the PUCs on this type of issue. Each utility was asked what types of items it leased, and the results are summarized in Table 7.1. Approximately half of those who knew about leasing arrangements indicated that their firm had none, or only a limited amount of leasing. Most of the remaining firms employed selected leasing, with transportation and high-technology items the most commonly mentioned. Only one firm mentioned any capital equipment items associated with power generation: gas turbines used for peaking units.

Two firms indicated that they used leasing extensively. One noted that virtually any significant item considered by the utility was subjected to a lease versus purchase analysis by a financial analyst in the

Table 7.1 - Utility Leasing Experience

None/negligible	7
Selected items:	7
High-technology (e.g., computers)	3
Capital equipment (gas turbines)	1
Land	2
Nuclear fuel	3
Office equipment	1
Numerous items	2
No answer/Don't know	6
Pollution control	0

firm. The second utility spokesman indicated that there had been "a heavy swing" toward leasing over the past few years, and that the firm now leased buildings, furniture, computers, typewriters, automobiles, etc. He noted that they had even talked to a major architect/engineering firm about leasing power plant generating equipment; raising capital was viewed as a problem, and any approaches such as ERC leasing that could be shown to be financially feasible would be considered. As one spokesman noted, "We have to be pretty creative about dollars these days."

None of the firms had leased pollution control equipment. One utility considered leasing a flue gas injection system for a limited period because it was undecided about the need for the system. A second had installed an FGD system with a regenerative process, and was in the process of selling the regeneration system and then leasing it back.

PROBLEMS WITH ERC LEASING

While a number of utilities indicated that they might be interested in leasing ERCs, several problems were identified. First (and most frequently mentioned) were the time conflicts between the position of lessor and lessee. The lessor would like to have relatively short-run access to the ERC (the reason for leasing it in the first place), while the lessee had a desire for longer-term security (for example, for the remaining life of an older plant scheduled for completion). But there could be competing pressures, as well. If the lessor was financing a scrubber, for example, he might want to have a firm income accounting position and hence a longer-term commitment on the lease. The lessee's purview was probably considerably less than the life of the asset, and this would result in the desire for a somewhat shorter commitment. The negotiation of acceptable agreements between the two parties could therefore become a major obstacle in ERC leasing transactions.

Second, many felt that the level of control required would be sufficiently stringent that utilities would control just to the emission reduction required. No overcontrol would be available. Many took the viewpoint expressed by representatives that "ERCs are going to get pretty expensive if an acid rain bill goes through . . . There won't be much excess available for leasing." They felt that the risk associated with generating ERCs was too high, given these high costs. One noted: "Utilities will only put on control up to the point where they have to." Basically this concern strikes at the heart of the market approach, which assumes that if the potential for return is sufficiently high, ERCs will be provided. The concept of ERC leasing is designed to reduce some of the uncertainty in the market. Apparently many utilities believe that leasing alone is not sufficient to overcome that risk.

This position is strengthened by comments in which the lessor is still viewed as having given up something valuable: "Those who have something [ERCs] will want to keep it." Several firms mentioned that any emissions reductions generated by overcontrol would be kept as a "cushion." "If the regs make us go down to 0.8 (lbs. SO_2 per MM BTU) and a scrubber gets to 0.6, we won't put the difference on the market. We'll probably need it because of reliability problems with the scrubbers." "If the thing [scrubber] hiccups or goes out of service for a short time, the difference is eaten up very quickly." As noted above, technical issues such as scrubber variability and outages were raised during the interviews, often with quite different perspectives. One utility justified a proposed bubble by noting the narrow range of scrubber variability, while another suggested that problems in scrubber operations (including reliability, variability, etc.,) would justify the need for a cushion. It is apparent, however, that the manner in which these technical issues concerning scrubber operations are finalized in ETP transactions could very well affect the availability of ERCs in an acid deposition trading program.

Similar attention focused on the question of demand. Many did not see demand for leased ERCs beyond those cases where a firm had an older unit it had decided to retire in the near future, and wanted to keep on line for a relatively short period of time. One firm noted that power generating equipment and its associated pollution control equipment were appreciating, and were a relatively good investment; the firm stated that it would not be particularly interested in leasing a portion of control equipment, but would probably make whatever investments were necessary for its own plants.

A fourth problem mentioned was the difficulty in setting up a legally binding agreement satisfactory to both leasing parties (lessee and lessor) and the regulatory bodies (the federal EPA, the state air pollution agency, and the Public Utility Commission). One utility representative suggested that the lease would require the signatures of at least three parties (lessee, lessor, and state air pollution agency). As noted earlier, problems in negotiating terms between the two utilities could become a major obstacle; the introduction of restrictive terms by other parties could become a severe impediment.

A number of the firms stated that the PUC would become a factor in ERC leasing, particularly since it would be a new issue and the sums spent in the leasing could be substantial. The intangible nature of ERCs and the potential impact on "used and useful" criterion in the rate base were identified as potential problems in PUC rulings.

A final impediment was attitudinal. One person commented on the issue in much the same terms as scrubbers were discussed in the first section. "We're in the electricity selling business, not in the air quality increment leasing business." Another stated "In all of our leases, we've always been the lessee—never the lessor." The unfamiliarity with leasing evinced by many firms, in conjunction with the novelty of ERC leasing, would probably present problems in implementation. Indeed, most of the firms still considered the ETP itself a novelty.

The novelty of ERC leasing also led to one particular type of sampling problem in this question not encountered in the others in the survey. Since most persons had not considered the issue, their viewpoints often changed over the course of the discussion. For example, one spokesman at first maintained that ERC leasing would not work because "those who have something [ERCs] will want to keep it." The discussion then focused on what would be "lost" in the leasing transaction if the utility ultimately maintained control of the ERC. Upon reflection, this spokesman shifted his response, indicating that perhaps leasing could work in certain situations after all. Another spokesman who had an identical initial reaction, however, remained convinced that anyone who held an ERC would keep it despite leasing options.

PUBLIC UTILITY COMMISSIONS

Most PUC representatives indicated that leasing had seldom been used by electric utilities in their state. A number noted that small items and technical items with rapid obsolescence periods were leased; none was aware of any major generating equipment leases, with two exceptions: (a) the combustion turbine described earlier; and (b) a leveraged lease for equipment associated with a coal conversion.

Leveraged leases are described in Chapter 12. While many industries shifted toward the use of leveraged leases in the 1970s, this was the only case of a leveraged lease found in the survey. No PUCs were aware of any leasing for pollution control equipment. When asked about the future potential for leasing (including ERCs), two PUCs responded favorably, stating "We'd be favorable to leasing, but there should not be a push toward flexibility if it hurts air quality." "Utilities have not been able to justify it [leasing], but the [PUC] would be receptive if conditions change." The remainder either noted that each case would have to be handled on its individual merit, or declined to take a position.

When PUCs were asked about impediments to ERC leasing, particularly in the rate-making area, only a few responses were offered. The initial reaction by a number of PUCs was that leased pollution control equipment or leased ERCs would be treated as operating expenses, rather than going into the rate base for depreciation. The results of either method would be essentially the same as far as ratepayers are concerned. One PUC member also raised the issue about the treatment of utility income from the lease, but was not certain how it would be resolved.

8
Interstate Trading

ELECTRIC UTILITIES

The question about the feasibility of interstate trading drew the most spirited responses from utility persons interviewed, which included the only unprintable expletive. The Edison Electric Institute had termed interstate trading "totally inconsistent with political reality," and the survey offered utilities an opportunity to agree or disagree with this characterization. The results indicate that 70 percent of the responding utilities agreed with this viewpoint: No, 16; Yes, 3; Yes, but only as part of a systems bubble, 2; and No position/don't know, 2. The remaining firms took a more positive view on interstate trading, although many qualified their answer to some degree. Individual viewpoints are outlined in greater detail in the following sections.

Negative Responses

The initial response of those who believed interstate trading was not feasible was often quite emphatic: "No way." "Interstate trading is unrealistic." "This is just a dream."

Several reasons for these viewpoints were given. The first and most common concern was competition between states. The competition was seen to exist on several levels—for protection of jobs and local firms, the state's coal industry, and future economic growth. Since a program to control acid deposition would include billions of dollars of capital equipment and construction, individual states would want to ensure that ratepayer's money provided local construction workers and suppliers with work. This might not happen if an in-state utility purchased ERCs from an out-of-state utility, and thus presented an obstacle to interstate trading.

Protection of the local coal industry has already been discussed in several of the cost analyses through "local coal" provisions. But utilities thought that the local coal issue and interstate trading would be linked—politically, if in no other way. For example, one firm noted that the Section 125 provisions of the Clean Air Act concerning "regionally available" coal had already been raised in Ohio, but eventually the issue "died out." "We're concerned, though, that the issue will arise again in the context of acid rain." While the firm stated that it disagreed with defining "regionally available" on the basis of a single state alone, this proposal had its proponents in the state. Any control mechanism containing interstate provisions (for example, trading) would, at a minimum, be scrutinized for its "local coal" impact. More likely, parochial interests such as local coal provisions would effectively remove the opportunity for any interstate trading at all.

As noted in previous sections, utilities have an incentive to maintain ERCs for future customers in the area. Since an ERC could ultimately affect the economic development of an area, utilities noted that their attempts to sell the ERC to an out-of-state utility would be met with resistance by state officials desiring to maintain the in-state potential for economic growth.

While this might be a potential impediment, the situation could be different in different states, depending on the extent to which industry is required to control in an acid deposition program. In particular, the potential new source review procedures for industrial sources require attention. It could be hypothesized, for example, that acid deposition controls would fall solely on the utility sector in a given state. If the utility installed a FGD system to meet its control requirement, its localized impact would be reduced. This might afford new industry a better chance of obtaining a new source permit, whether or not the utility sold a portion of its reduction as an ERC. On the other hand, if new industry were subject to acid deposition offset requirements as well, the potential sale of ERCs could very well affect its development. A number of firms noted the disagreements that have arisen between states over PSD increment consumption, and indicated that economic growth issues in interstate trading would be comparable.

Differences between states were not limited to economic issues. Several utilities cited the legal battles underway between states under Section 126 of the Clean Air Act. This portion of the act allows the EPA Administrator to act if an emission source in one state contributes to the violation of air quality standards in another. Wetstone and Ostrov have discussed the potential usefulness of this section for an acid deposition control program.[1] But while Section 126 has been ineffective for acid deposition control, it has nonetheless generated some animosity. One utility spokesperson said: "We've been sued by ——— under Section 126, so I don't think there will be a favorable response to interstate trading there." Another utility cited two state air pollution agencies that "don't get along all that well" as a drawback to interstate trading; he claimed that the problems were based in part on Section 126 disagreements.

Other impediments cited were site-specific or problematical. A municipally owned utility noted that it would be difficult for it to use municipal funding to purchase an ERC in another state. To some extent, the issue of an ERC as an intangible asset applied here, since the utility was certainly willing to purchase materials and equipment from out-of-state firms. It also noted that it would not sell any ERCs generated but would retain them for future industrial development in the municipality.

Other firms noted that legal and tax concerns about ERCs could vary from state to state, and this might make agreements more difficult to implement. Another suggested that commerce laws might require consideration. Kontnik, in an article addressing the legal implications of emission trading, noted that 49 states were affected by:

> The comprehensive and complex provisions of Article 2 of the Uniform Commercial Code [UCC] [which] automatically apply to "transactions in (usually sales of) goods" . . . Although control program agreements are not likely to be covered by Article 2, the sale of emission reduction credits (and less likely, the lease) might well be . . . Whether and how Article 2 applies will not be known until courts consider the question.[2]

One utility lawyer noted that, while the Supreme Court had prohibited commerce laws between the states, states had done quite well in devising de facto trade barriers. The extent to which legal mechanisms could be used to thwart interstate trading by persons opposed to the transaction was not known.

Given the potential impediments and uncertainties, some felt that the amount of effort necessary to negotiate an agreement between two utilities, to overcome legal, political, or other impediments described above, and then to obtain approval from all affected state agencies in both states (and potentially the federal government) was simply not worth while. As two spokespersons stated: "Interstate trading will be complicated, and it's doubtful whether the savings will be worth it." "There simply isn't sufficient incentive to trade out of state." Finally, one person characterized interstate trading as nothing more than a "political sop"; he contended that no one believed that it would work, but that it could be useful in the debate concerning acid deposition control as a bargaining chip—one that would readily be discarded.

Interstate Systems Bubbles

Two utilities indicated that interstate trading would be feasible, but only in one situation—where the interstate trade was part of a systems bubble. Both firms had generating plants in more than one state, so the potential existed in each case. Both also agreed that the number of firms able to use the mechanism would be sufficiently small that interstate trading would not be a significant factor in an acid deposition control program. A third firm in a comparable situation (but contacted by one of the PUCs) also had a positive response.

Another utility in a similar situation stated that interstate trading would not work, even as part of a systems bubble. The spokesman noted: "If the plant in one state is made tighter to allow for emissions from a plant in another state, the state controlling greater will scream." There might be some way of distributing the costs between the two plants, but this may bring about other problems with the PUC, complaints from ratepayers, and so on.

Another firm, that had responded negatively to interstate emissions trading, indicated that there might be some opportunity for trade if the emphasis were placed on utility firms in one state trading with industrial firms in another, instead of utility-utility trades. He suggested that removing the involvement of one of the PUCs could make trades considerably easier to implement.

Positive Responses

Three firms thought that interstate emissions trading was feasible, although one response was qualified. The first utility representative indicated that he could probably get his management to agree to an interstate trade if he could show that it was cost-effective and politically feasible. While he was convinced that he could probably document the former, "there are a lot of questions about the latter." The decision to proceed with an interstate trade would probably hinge on the reception given to a preliminary proposal.

A spokesman for a second firm indicated that he could see no reasons why interstate trades would not be feasible. He indicated: "We'd be willing to explore whatever options are cheapest, including trades out of state." When told that this viewpoint was contrary to the EEI position, he remained firm: "All of the details would have to be addressed, but it could still work." A third firm's spokesman stated that "eventually there will be interstate trading." He suggested that the groundwork should be laid now within the existing ETP for trades in interstate metropolitan areas or regions (e.g., Youngstown-Pittsburgh or Ashtabula-Erie). He concluded that "there is no reason *not* to trade."

PUBLIC UTILITY COMMISSIONS

"We *are* the problem" was the way that one PUC representative characterized the feasibility of interstate emissions trading. He stated that "the utilities are probably right" in suggesting that interstate trading was inconsistent with political reality. What makes these comments particularly significant is the fact that the spokesman represented one of the states in Table 5.1 with significant reduction requirements. He indicated that the PUC would probably not allow utilities to overcontrol (even if the utility in another state paid for it). The PUC would probably not allow utilities to buy ERCs originating in another state. He explained that he recognized that interstate trading made economic sense and, given the fact that his state had a large control requirement, also recognized that it would probably benefit more than most others. Despite these factors however, interstate trading did not make "political sense," given the concern over jobs, local coal, and the potential for misunderstanding the issue. Headlines reading "Ohio Utilities Overcontrolling Emissions," if the ERC were to be created and sold, or "Ohio Ratepayers Paying for Scrubber in Pennsylvania," if it were to be purchased, could occur. While the technical-economic case could be considered valid, it would merely result in "one more highly technical area moved into the public purview," with the attendant potential for misunderstanding and political manipulations. The spokesman indicated that the PUC was "highly politicized, probably now more than ever." These obstacles for interstate trading could probably be overcome if all of the utilities in the state put on a "united PR effort" to develop the potential for trading, but he didn't believe that the utilities were likely to do so.

Not all the PUC responses were as negative. The question about potential headlines was raised with several other PUCs with significant control requirements. Several PUCs indicated that the first headline, concerning ERC purchases from out-of-state, would not cause serious problems. Some were relatively unconcerned, noting "Its really not much different than spending ratepayer's money in Montana or Wyoming for coal." Others were more cautious: "It could lead to problems, but it probably wouldn't destroy the trade" and "It wouldn't necessarily be a problem if an adequate analysis of the options had been performed."

They were more concerned about a headline that might read "Ohio Utility Selling a Growth Increment to Pennsylvania." The growth issue was more serious: "This is a more loaded question politically." "We couldn't ignore the implications of the growth issue and the impact on the recovery of [the state's] economy." A number of states indicated that, while growth is an important issue, it would be considered as part of the review of the air pollution control agency. Two PUCs noted that growth concerns about increment consumption were handled in PSD permits by the air agency. On the other hand, another PUC representative noted that while his PUC did not normally get involved in environmental issues, it would do so "if there was a question about saving ERCs for future growth in the state."

One PUC noted that if one of their utilities purchased ERCs from an out-of-state firm, then "we would not be getting the localized emissions reductions in this case—making it more difficult to make the case." Two other PUCs thought that interstate trading would be feasible if the scope were limited somewhat. One suggested that trades be limited to contiguous states. The other suggested that they had already "kept a pretty close eye on generating plants operating outside the state which would affect the state." (The plants he gave as examples would be eligible under an interstate systems bubble.)

9

Implications of Survey Results for the Market

The survey of electric utilities concerning emissions trading confirmed a strong bias toward internal transactions—not only in the case histories, but also in the viewpoint of those interviewed. The term "EPA's Emissions Trading Program" connoted to most utility representatives the use of offsets and emissions banking, typically associated with external trades. Bubbles, on the other hand, are now viewed as merely a flexible extension of the permitting process, and virtually all of the discussion about bubbles focused on internal transactions.

The infrastructure for emissions trading by utilities is apparently well established. Every utility representative was familiar with emissions trading and its components, and virtually all had evaluated the potential use of emissions trading at generating stations in their system.

Nonetheless, there was a surprising lack of formal experience with the ETP. Utilities seem content to make what might be termed "passive" use of emissions trading—keeping up to date with regulatory opportunities, evaluating internal options, even banking retired units—but are reluctant to take the steps for "active" participation (for example, banking overcontrol or external trades).

Reasons for the slow development of the market were discussed in Chapter 3. During the survey, it became apparent that the general impediments (market constraints, uncertainty, transaction costs, and others such as economic conditions) were more important factors than most utility-specific impediments (rate making risks, maintaining ERCs for customers, and long planning times). The one utility-specific impediment that had played an important role was the enhanced potential for regulatory requirements—that is, the likelihood of an acid deposition control program itself. Emissions trading in an acid deposition control program is likely to show characteristics similar to the ETP. There was strong utility support for systems bubbles: indeed, many predicted problems if this type of emissions trading were *not* allowed.

External trading, whether intra- or interstate, generated much less enthusiasm.

For external trading to work in an acid deposition context, it will be necessary to have utilities generate a sufficient quantity of ERCs, and then be willing to use the ERCs in the market. The survey found cases where potential ERCs were generated, and the technical feasibility of ERC generation through unit retirement or overcontrol seems assured. But while the utilities are willing to view unit retirements as ERCs (but maintaining a reluctance to sell them), they indicated that there was not similar interest for overcontrol. Instead, overcontrol was viewed as a cushion, important for negotiations with state air pollution agencies. It has not made significant inroads in the ETP for this reason and two others: (a) a perception that the overcontrol could be easily "lost" or made mandatory, and (b) an apparent reluctance on the part of state agencies to give ERCs for overcontrol. If external trades are to work, these problems must be considered.

Utilities will probably maintain a cushion to some extent, regardless of the regulations—indeed, this has a certain environmental benefit. But to avoid non-economic overcontrol, it would be worthwhile to define clearly how upsets, outages, malfunctions, etc., associated with FGD operations will affect an ERC. In effect, it will indicate to utilities how great a cushion is necessary. A comparable situation exists for handling coal sulfur variability if overcontrol is to be achieved through low-sulfur coal.

The fear of losing the overcontrol or having it made mandatory will always exist, particularly given the uncertainty concerning the impact of emissions reductions to mitigate acid deposition. While future regulations are always a possibility, the major problem seems to be a fear of spending funds and getting no credit for the actions (and the possibility of having to spend additional funds). While the utilities' experience with overcontrol in the past has not always been positive, the ETP offered a means of formalizing credit for the action. Thus formalizing the credit as early in the process as possible, and assisting in developing a proper means for using the credit (through leasing, for example) should help. Nonetheless, the problem is basically one of attitude based on prior negative experience. The ETP is potentially a solution to this problem, not an impediment.

Finally, the viewpoint of the states in giving credit for overcontrol is also attitudinal. As states can take a more stringent approach than EPA, there appears to be little that EPA could do in this area other than to foster awareness of the problem and the potential impact it could have on emissions trading for acid deposition.

Assuming that utilities are willing to generate ERCs from overcontrol or unit retirement, they will still be reluctant to sell the ERCs. The survey asked firms if ERC leasing arrangements would improve use of the market, and the response was that it would work in a certain number of situations (for example, to continue operations on an older unit nearing retirement). A number of potential leasing cases were identified in the course of the survey even without an acid deposition control requirement, so there does appear to be some merit to this approach. Problems were identified, however, including time conflicts between the lessor and the lessee, utilities' lack of experience with leasing in the past, and attitudinal concerns.

Part III

◆

The Need for ERC Leasing

10
Methodology

The analysis of emissions trading by electric utilities outlined in Part One found that these firms were not necessarily cost minimizers, but were guided by the strategic aspects of the control option. The utilities face a number of impediments that encourage them to hoard ERCs (such as the potential for future regulatory requirements and the protection of growth allowance for future customers). It was suggested that leasing arrangements that allowed electric utilities to maintain control while giving other firms use of the ERCs were likely to be important, particularly if large-scale market transactions by electric utilities were anticipated (for example, in an acid deposition control program).

The survey of utilities in the Midwest and Northeast outlined in Part Two was encouraging, even without any acid deposition requirement. For example, one utility had installed a scrubber to meet a state SIP requirement that could not be met with low-sulfur coal, and the design resulted in overcontrol. The utility was approached by another source who wanted to "borrow" the potential ERC for a limited time—an ideal candidate for a lease. In a second case, the utility approached a number of industrial firms seeking ERCs; one of the firms offered to lease the ERCs, but the utility was subsequently able to arrange for the purchase of other available ERCs.

These examples indicate that ERC leasing arrangements are certainly feasible, and a number of other studies have suggested that ERC leasing transactions deserve additional attention in the Emissions Trading Program. A 1982 ERC tax analysis by Baker and Winslow found that amortization deductions were likely to be denied for the outright purchase of an ERC.[1] A leasing transaction, on the other hand, allowed the ERC user to claim a current deduction for its rental payments. Baker and Winslow stated that "many firms may decide upon long term ERC leasing arrangements, if they are permitted by state regulations" in order to avoid the amortization problem.[2] They concluded that leasing options should be considered when emissions trading regulations are developed.

Similarly, Gerard studied the effects of public utility regulations on the efficiency of a market for emissions permits (that is, a full-scale market approach) and determined that the most reasonable means to introduce an emissions market system for the utility sector employed a leasing type of approach.[3]

The utility representatives in the survey raised a number of questions, however, concerning the potential supply of, as well as the demand for, leased ERCs (see Chapter 7). This section describes an analysis of the potential for leased ERCs under three different types of acid deposition regulatory control programs, and addresses both economic and implementation concerns.

The methodological approach to the leasing analysis was shaped by several considerations. First, and perhaps most important, we wanted to examine the feasibility in a "real world" setting. This implied that we use empirical utility emissions and cost data for the analysis, rather than hypothetical data. Second, we wanted the analysis to focus on an area where market transactions were likely to be substantive. Chapter 5 had identified four "core" states—Ohio, Pennsylvania, Illinois and Indiana—with the highest rankings concerning the quantity of utility-generated SO_2 emissions, the greatest coal-fired capacity, and the most significant potential SO_2 reduction requirements, so attention was focused in this area. Third, we were interested in obtaining a diverse mix of both utilities and unit sizes, so that the feasibility of leasing could be examined for all types of sources. Finally, other factors such as the resources available for the analysis and the cooperation of personnel in the previous contacts were also considered.

From these considerations it was determined that a statewide analysis would be employed, using Illinois as the target state. Illinois seems uniquely qualified: the analysis included individual boiler units ranging from 11 mW net to 840 mW net, with utility systems ranging from the same 11 mW net to approximately 16,900 mW net. Coal-fired capacity was available from major investor-owned utilities, rural power cooperatives, and municipal units. Sixty-two existing units and four units scheduled to come on-line in the 1990s were included in the study.

The choice of a statewide analysis eliminated the consideration of interstate ERC leasing transactions. The political difficulty of any type of interstate trading (regardless of its technical merit) has already been discussed, but an initial focus on the intrastate market is an appropriate first step in any event. Similarly, only utilities have been included in the study. Utility-industrial ERC leasing transactions may ultimately prove to be feasible, but this analysis is more narrowly targeted; it seems likely that, in any event, utility firms will generate the greatest quantity of low-cost ERCs.

OVERVIEW

The methodology employed in this analysis can be summarized in six basic steps:

1. Determine the marginal costs and effectiveness of sulfur dioxide control options available for each utility system in the state.
2. Determine the aggregate marginal costs and effectiveness for all these sulfur dioxide control options on a statewide basis.
3. Determine the emissions levels that would be allowed under various acid deposition regulatory control programs for individual utilities in the state.
4. Determine the appropriate economically efficient control technology to be employed by each utility to meet aggregate emissions totals in the state for each regulatory program.
5. Determine the nature of the resulting market in ERCs for each regulatory program.
6. Examine the potential for ERC leasing transactions in this market for each regulatory program, based upon projected emission levels, the potential exposure in future years if ERCs are sold, and the potential for decreasing demand in future years conducive to leasing.

These steps are discussed in greater detail below.

THE SEROP MODEL

The first and second steps in the methodology, which calculate both the individual utility and statewide aggregate marginal cost and effectiveness of the sulfur dioxide control options, employed a computer program known as SEROP. This program, the *S*ulfur Dioxide *E*mission *R*eduction *O*ptimization *P*rogram, was developed by systems engineers at the Wisconsin Power and Light Company in 1982 to analyze potential acid deposition control options for individual utilities.[4] The program uses operational and cost data for each control option, along with the projected annual generation and thermal performance of each unit, to calculate year by year marginal cost rankings for control options.

Figure 10.1 shows the general program logic and information flows. A number of SEROP features are important in defining the scope of the analysis. First, the model employs system bubbles to meet emission targets. It is thus assumed that trading between generating stations in a utility system will be allowed; existing trading focuses on localized air

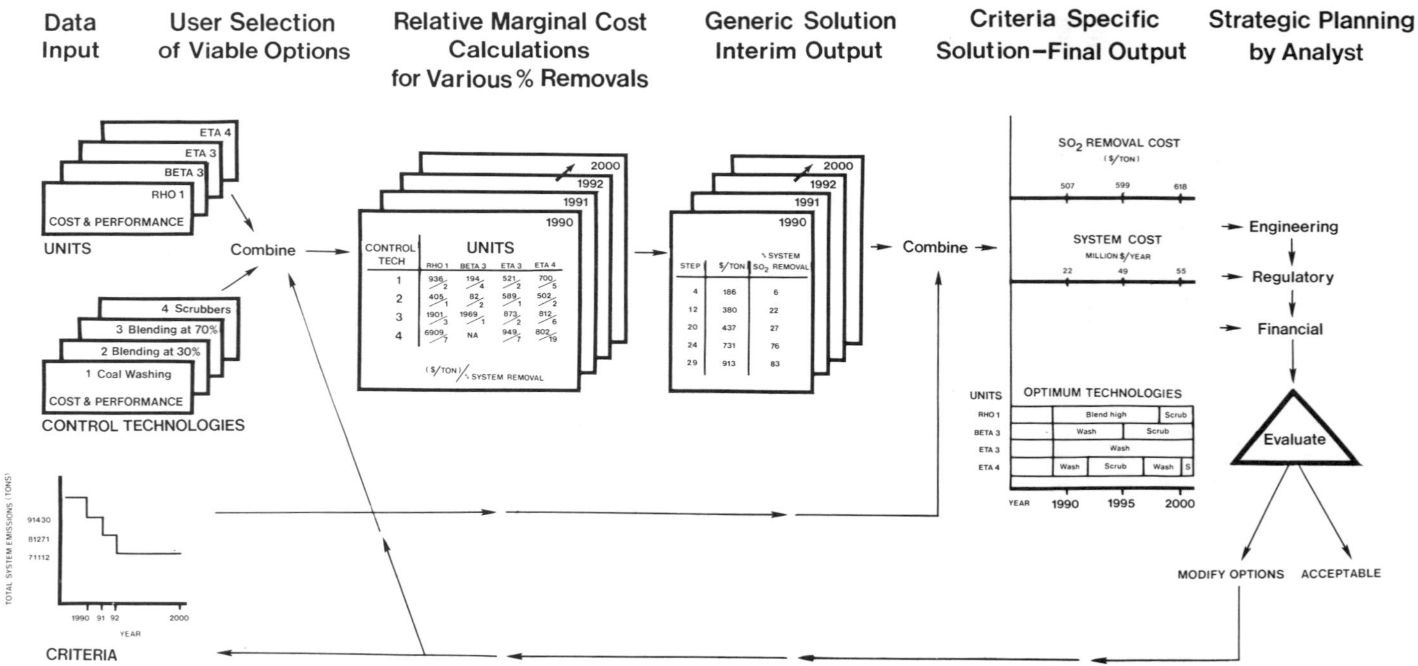

Figure 10.1 Sample SO_2 Emission Reduction Optimization Program (SEROP) Logic Diagram

Source: D. G. Neumeyer, "Analyzing Emission Controls for Power Plants," *Pollution Engineering* 15, no. 4 (April 1983). © *Pollution Engineering;* reprinted by permission.

quality concentrations and is therefore more limited in a geographic sense. As Neumeyer notes, the model assumes "that emission removal anywhere within a region is effective in the reduction of any long range transport situation."[5]

Second, the model allows the user to select up to a 30-year period for analysis. Given the limitations of the production cost modeling estimates and demand forecasting, and the targeted period for emissions reductions in proposed legislation, the period 1983 to 2000 was chosen for this analysis. The model compares all control options within a given year, and the economically efficient control technologies were selected based on the relative rankings in those years. This limits intertemporal comparisons of control options, and also gives equivalent weighting to early years in the study (1983, 1984, etc.) even though control was not or would not be implemented. Obviously, these simplifying assumptions would not be made by utilities assessing their own response to a control requirement, and thus represent a limitation of the study. We recognize that a control program might not be implemented until the 1990s, and that the relative timing of the costs for various options for multiple utilities will play a more important role than SEROP can realistically evaluate. The more favorable leasing transactions identified in this analysis were subsequently examined with a more sophisticated revenue requirement model to account for alternative tax treatments and time dependent factors (such as discount rate and inflation), but numerous other utility options, such as shifts in the generation mix, bulk power purchase agreements, and load management, were not considered.

Finally, SEROP was designed for individual utility systems, and each of the ten utilities in the state was analyzed individually. The model was also used to develop aggregate marginal cost data, however, with all of the utilities in the entire state collected into one large computer run. The analysis assumed a fixed carrying charge of 18.2 percent for all ten utilities. The 18.2 percent estimate is reasonably close to the levels reported by the two Illinois utilities in an Edison Electric Institute survey (18.25 percent and 19.05 percent),[6] and the example analyses of both EEI and SEROP (18.2 and 18.1 percent, respectively).[7] To the extent that these figures vary from utility to utility, however, the relative ranking of control options may vary slightly.

SULFUR DIOXIDE CONTROL OPTIONS

SEROP allows the user to evaluate numerous control options (up to 100) for each individual unit. Seven control options were evaluated in this analysis, as follows:

1. Coal washing,
2. Blending to 3.0‰ SO_2 per MM BTU heat input,
3. Blending to 1.8‰ SO_2 per MM BTU heat input,
4. Switching to low-sulfur coal,
5. Flue gas desulfurization (utilizing wet limestone scrubbers),
6. FGD on low-sulfur coal, and
7. Overcontrol on existing FGD systems.

The first option was open to surprisingly few firms, since the majority already obtain washed coal. In blending and low-sulfur coal options, the cost and coal quality of low-sulfur supplies already arranged by the utility were employed whenever possible. Assumptions for sizing the FGD units are described below. It was assumed that any unit currently burning low-sulfur coal would not shift back to higher-sulfur coal; its only control option would be to scrub the low-sulfur fuel. Finally, a number of FGD systems already in operation in the state are not operating at their design control efficiency (although one is currently surpassing this level), because these units are able to meet emission requirements at lower levels. Operation at the design efficiency was analyzed as the seventh control option.

The development of cost data for these options, as well as operational data concerning the individual units, is discussed in detail below.

ACID DEPOSITION CONTROL SCENARIOS

When the SEROP modeling analysis in the first two steps had been completed, the third step determined allowable emission levels under each acid deposition regulatory control program for each utility. Three types of regulatory control programs were evaluated:

1. The 1.5 lbs. Excess Emission Reduction. Under this type of control program, each unit was evaluated to determine whether it emitted more or less than 1.5 lbs. SO_2 per MM BTU heat input in the base year (1983). If the unit emitted less than 1.5 lbs. SO_2 per MM BTU, then no emissions reductions were assessed. If it emitted more than 1.5 lbs. SO_2 per MM BTU, the emissions greater than this level were calculated, and these "excess" emissions over the system were summarized. This represented the emission reduction required in the base year. The allowable level was then determined, and this was held constant for all subsequent years. This approach is similar to the one employed in the EEI survey and in S-3041, although the base years are different. It allows no credit for those units emitting below the target level.

2. A proportional 50 percent reduction. As the name implies, total

emissions in the base year are determined for each utility, and a 50 percent reduction is required. The allowable emission level (that is, 50 percent of emissions in the base year) is held constant in subsequent years. This strategy obviously penalizes utilities that have already made substantial control efforts, and would therefore face stiff political resistance.

3. A 1.8 lbs. SO_2 per MM BTU heat input systems cap. This strategy would require that utilities emitting more than 1.8 lbs. SO2 per MM BTU on a systemwide basis in the base year reduce their emissions to this level. This emission limitation would then be held constant in subsequent years. Note that this does *not* allow a 1.8 lbs. SO_2 per MM BTU level in all subsequent years, since emissions could increase with growth over time; the resulting program could therefore require system levels below 1.8 lbs. SO_2 per MM BTU in future years. Similarly, utility systems that have a system average below 1.8 lbs. SO_2 per MM BTU in the base year would be required to maintain that emission limitation in subsequent years; they would not be allowed to increase emissions to the 1.8 level.

OPERATION OF THE MARKET

The first three steps of the methodology have thus far determined individual marginal cost curves for each utility, the aggregate marginal cost curve for the state, and the allowable emission levels under various acid deposition control strategies. The fourth and fifth steps use this information to determine the individual response of each utility, and to determine the nature of the market in ERCs.

Figure 10.2 shows how the market would operate for two firms, with individual marginal cost curves MC_1 and MC_2 and an aggregate marginal cost curve MC_{1+2}. The allowable emissions would require emission reductions at the levels marked X_1 and X_2. If Utility 1 were to ignore the market, its marginal cost at the allowable level would be P_1, and the total cost of control would be the area under MC_1 up to X_1. Utility 2, on the other hand, can meet its allowable emission level at a much lower marginal cost (p_2).

The cost-efficient approach, however, is to meet the total allowable level (X_1 + X_2) at *g*, where the marginal cost is P_{1+2}. Under these conditions, Utility 1 would control only up to *a*, and purchase the remaining necessary ERCs (*ac*) from Utility 2. Utility 2 would control beyond its required level to *e*, and sell the additional ERCs (*de* = *ac*) to Utility 1.

Utility 1 would make a transfer payment equal to the aggregate marginal cost times the number of ERCs (that is, (P_{1+2}) × (*ac*)), which

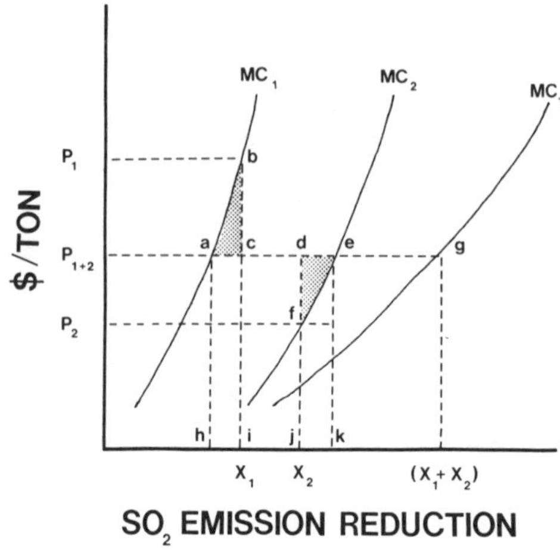

Figure 10.2 Operation of the Market

is the rectangular area *haci* in Figure 10.2. Note that this is less than the cost of control if Utility 1 decided to ignore the market, because that amount of control would cost *habci*. Thus, Utility 1 saves the shaded area *abc* by paying Utility 2 for the ERCs.

Utility 2 receives the transfer payment *haci*, which is equal to *jdek*. However, the cost of providing the ERCs is the area under the MC_2 cost curve between *j* and *k*. Utility 2 must therefore spend only *jfek*, and can save the remaining shaded area *fde*.

Thus, the total cost of SO_2 control under the cost-efficient case is the area under the MC_{1+2} curve up to $X_1 + X_2$. The savings equal the sum of the shaded areas, which represent the gain of both utilities from trading over nontrading control strategies. The transfer payments equal the aggregate marginal price times the number of ERCs bought or sold (note that this must not be counted twice).

The analysis outlined in the following sections considered ten utilities instead of two, but the approach is identical for determining the level of control technology employed by each utility, the quantities of ERCs bought and sold, the market price for ERCs, etc.

Finally, when the market was defined, it was possible to take the sixth step and examine the potential for ERC leasing transactions. This was done by considering factors such as the size of the transaction by each utility in each year, the need for ERCs in future years, and the potential for the utility to provide its ERCs internally. (More detailed aspects of this and other steps in the methodology are discussed in the next chapter.)

DATA BASE DEVELOPMENT

The most recent DOE inventory of steam electric generating stations at the time of the analysis listed sixteen utilities in Illinois with coal-fired capacity.[8] These included five major privately owned systems, two rural cooperatives, and nine municipalities.

To collect necessary data for the modeling analysis, personal interviews were obtained with the four largest investor-owned utilities as well as with several of the smaller municipal units. The remaining utilities were contacted by telephone. We soon learned that the DOE inventory data was somewhat dated. Some of the municipal units have not burned coal for more than ten years, several had retired their units, and at least one had been dismantled. Six municipalities were eliminated in this manner, leaving ten utilities in the analysis. As outlined in Table 10.1, both utility and individual unit sizes are quite diverse. The cooperative and municipal units were considerably smaller than those that were part of the privately owned systems; indeed, the five municipal and cooperative utilities combined accounted for less than 5 percent of the coal-fired capacity in the state. Nonetheless, these utilities were treated individually throughout the analysis.

The data sought from the utilities in the personal visits and telephone interviews can be divided into two general categories: operational data and cost data.

Table 10.1 - Illinois Coal-Fired Capacity (MWnet)

	Utilities			Unit		
Size	P	M	C	P	M	C
< 50 MW		2	1	1	4	4
50-100				8	2	
100-250				16	1	1
250-500		1	1	12		
500-1000				13		
1000-2500	2					
2500-5000	2					
> 5000	1					

Note:
P = privately owned, M = municipal, and C = cooperative.

Operational Data

The first stage of the interview confirmed basic inventory data such as unit size and initial year of operation, and also requested information concerning each unit's heat rate and the sulfur and heat content of the coal. Existing pollution control measures (for example, blending operations, low-sulfur coal modifications, and existing scrubber efficiencies) were discussed, and opinions concerning the technical feasibility of the seven potential control options were then sought.

Finally, production cost modeling projections to the year 2000 were requested, if available. Two of the utilities made the detailed unit-by-unit results available, and one took the trouble to make a special updated computer run in order to provide the most recent data available. Two other utilities provided total system load forecasts, and the units were allocated considering historical capacity factor data provided. One major private utility turned down the request for production cost modeling data, and its load was forecast using utility-specific data from the 1982 Illinois Energy Plan,[9] with the exception that a slightly lower rate of increase in net energy was employed (2.5 percent instead of 3.3 percent). The load was allocated using both historical capacity factors and updated information concerning the scheduled operation of new units currently under construction. The remaining utilities did not employ production cost modeling, and future net energy production and heat rates were estimated based on conversations with the utility representatives.

It is anticipated that four new coal-fired units will come on-line before the year 2000, and approximately six smaller units will be retired. These new and retired units were accounted for in the projections.

Data summarizing the unit rating, average annual heat rates, and power production for 1983 to 2000 for each of the units in the analysis are available.[10] Throughout the remainder of this report, each of the 66 units is identified by a five-character alphanumeric code (to preserve the confidentiality of the individual utilities). The first character (A through J) represents the utility. The second (A through H) represents a plant site. The third and fourth digits indicate the year that the unit was initially brought on line (for example, 76 = 1976). Since these four characters above were not sufficient (for example, two units were brought on-line in 1948 by utility B at plant site D), a fifth character was added after a hyphen. It numbers the units at the plant site and reaches as high as six (at DA55–6).

Cost Data

For each of the existing and potentially feasible control options, three types of cost were determined: (a) the capital investment (in $/kW); (b) the fixed production cost increase (in $/year per kW); and (c) the variable production cost increase (in $/mWh). These were developed using five general sources of information, in order of preference:

1. *Utility Estimates*. In a number of cases, the utilities have done detailed cost analyses of specific options, and these data were requested and utilized whenever possible. In addition, "generic" or "fallback" estimates for options such as scrubbers were developed from sources discussed below; the utilities were asked to comment on those estimates for their particular plants, to identify site-specific problems that would render them inaccurate, and other unique considerations.

2. *The Edison Electric Institute Survey Results*. Two Illinois utilities responded to the March 1983 EEI survey concerning acid rain control costs, and these responses were reviewed.[11] One of the utilities was questioned about its estimated FGD energy penalty, which seemed extremely high. A lower estimate (approximately 25 percent of the original value) was subsequently agreed upon. Certain other data were similarly revised or updated.

3. *ICF, Inc., Analyses*. ICF, Inc., has performed a number of regulatory impact analyses over recent years. One study that addressed the costs of S-3041 included a wealth of information concerning both new and retrofit FGD systems, upgrades for particulate control equipment for shifts to low-sulfur coal, future costs of low-sulfur coal, etc.[12] The results of this ICF cost analysis were used in the acid rain briefing document prepared for Ruckelshaus by EPA's Acid Deposition Task Force.[13] These cost estimates were therefore employed whenever utility-specific data were not available. A number of points should be made about these estimates. First, the cost of a specific control option for an individual unit does not strictly depend upon the unit itself—it also depends upon other units at the plant site. ICF outlines cost estimates for a 500 mW plant. It then notes that capital and fixed O&M costs are approximately 50 percent higher for a 250 mW plant, and 100 percent higher for a 100 mW plant. Estimates for the scrubber costs in the SEROP analysis take into account the total size of comparably controlled units at each plant site. For example, if a utility currently has two 200 mW units and one 100 mW unit (for a total of 500 mW) burning high-sulfur Illinois coal at a given plant site, the capital and fixed O&M costs input into the model are equivalent for all three units (on a $/kW and $/yr per kW basis), and all are at the 500 mW price. Should the utility choose to scrub only one or two of the three units, the cost estimates for each unit in the SEROP analysis would be low. Thus, if the 100 mW unit alone is scrubbed, the costs for this unit in the SEROP analysis would be low by approximately a factor of two. Similar "shared" costs exist for other cost

elements as well (such as some coal-handling equipment and railroad upgrade costs).

Second, particulate equipment upgrade costs in the ICF report are aggregate average costs (in $/kW) that assume that 75 percent of the firms install flue gas conditioning, and 25 percent add new ESP plate area. These data were nonetheless used to estimate costs for individual plants, for three reasons: (a) they provided detailed data for a wide range of both pre- and post-control coal shifts; (b) they provided estimates that agreed quite closely with the recent experience of at least one major utility in the state; and (c) in the cases where particulate control factors were likely to be of critical importance, separate control estimates were made (see below).

4. *Molburg-Rubin Cost Models*. In discussions with a number of the utility representatives, it became apparent that certain units would require more than an upgrade of existing particulate control equipment: new particulate control equipment would be necessary if a shift to lower-sulfur coal were made. For these units, site-specific cost estimates for fabric filtration control devices (baghouses) were obtained using the pollution control costing models developed by Molburg and Rubin of Carnegie-Mellon.[14] Data required for specific variables (for example, the air-to-cloth ratio and fabric filter bag costs) were obtained from the recent literature.[15]

Coal-washing costs were obtained from an Argonne National Laboratory analysis of coal washing in the Midwest.[16] Argonne similarly based their coal washing analysis on the Molburg-Rubin models.

5. *Other*. In those cases where the four sources listed above were not sufficient, other sources were employed. For example, units CD67-1 and CD68-2 would require both new railroad access and the upgrade of existing track to handle a switch to low-sulfur coal. These costs were estimated using construction, upgrading, land acquisition requirements, etc., for level terrain.[17] Similarly, the cost of developing a relatively sophisticated coal blending system was included for the plants with larger coal burning units.[18] Plants with smaller units were assumed to use bulldozers or similar unsophisticated blending procedures.

11
Operation of the Market

The first task included the development of utility-specific marginal data for sulfur dioxide emission reductions. These were developed for each year for each utility, and an example of the results for the base year (1983) are shown in Figure 11.1. The results clearly show the significant differences between utilities in both the potential quantity and cost of emissions reductions. (The three smallest utilities, F, I, and J, were omitted from this figure to keep the scales reasonable.) The aggregate statewide marginal cost curve for the same year is shown in Figure 11.2. This shows the general form one would expect, with the costs rising dramatically as SO_2 emissions reductions exceed 700,000 tons/year.

One must always take care, of course, when interpreting marginal control curves obtained from discrete technical options. In certain areas, the curves are nearly continuous; for example, instead of blending two coals to meet a 1.8 lbs. SO_2 level, one could just as easily target 1.79 or 1.81, with very small shifts in costs accompanying the slight shift in emissions reductions. On the other hand, at the point where the utility installs an FGD system, there will be a major shift in both cost and emissions reductions and the curve is not continuous in this region.

Twenty-nine control options are already being employed in the state. These options reduce SO_2 emissions over 800,000 tons below the levels that would be obtained if high-sulfur Illinois coal were burned in all coal-fired units in the state (without controls). The marginal cost data outlined in Figures 11.1 and 11.2 are based on control options beyond these 29 existing controls.

ALLOWABLE EMISSIONS FOR THE ACID DEPOSITION CONTROL PROGRAMS

The allowable emissions calculated under each of the three control programs discussed are outlined in Table 11.1. It can be seen that the 1.5 lbs. excess emissions reduction is the most stringent on a statewide

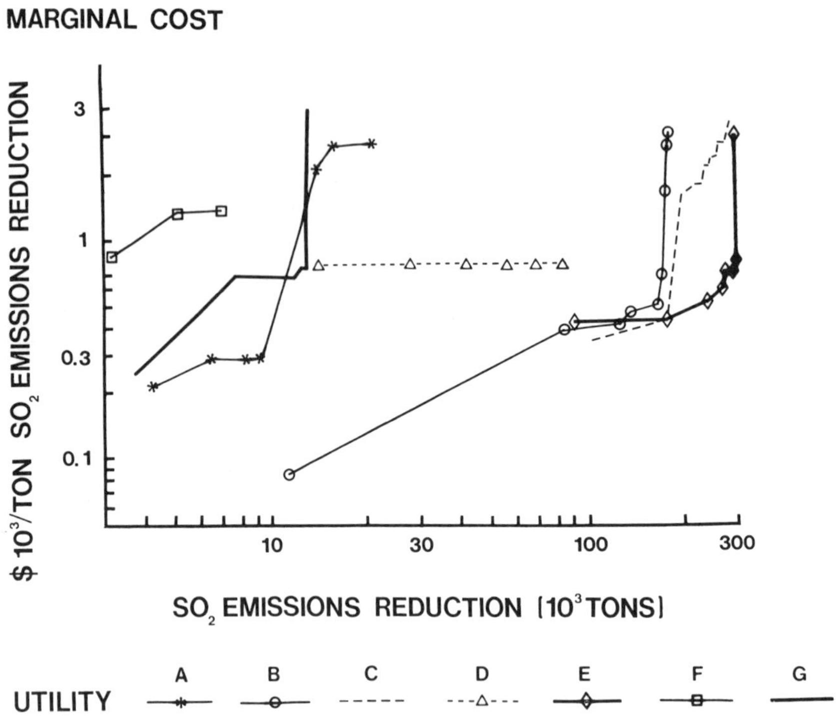

Figure 11.1 Utility-Specific Marginal Cost of Control (1983)

basis, followed by the 50 percent reduction and the 1.8 lbs. systems cap. The impacts vary considerably from utility to utility, however. Note that utility A would be heavily penalized under a 50 percent reduction because it already uses low-sulfur coal quite extensively, while utility B is essentially the opposite. The 1.8 system cap would aid utility C because it has numerous plants burning low-sulfur coal well below 1.8 lbs., as well as a major plant emitting above 6 lbs. SO_2. It would no doubt find this option the most preferable, and this has important results as far as the market analysis is concerned (to be discussed below).

ECONOMICALLY EFFICIENT CONTROL TECHNOLOGY

With the statewide aggregate marginal cost data, it is possible to determine the control technologies in the state that will lead to an economically efficient solution of the statewide emission levels manda-

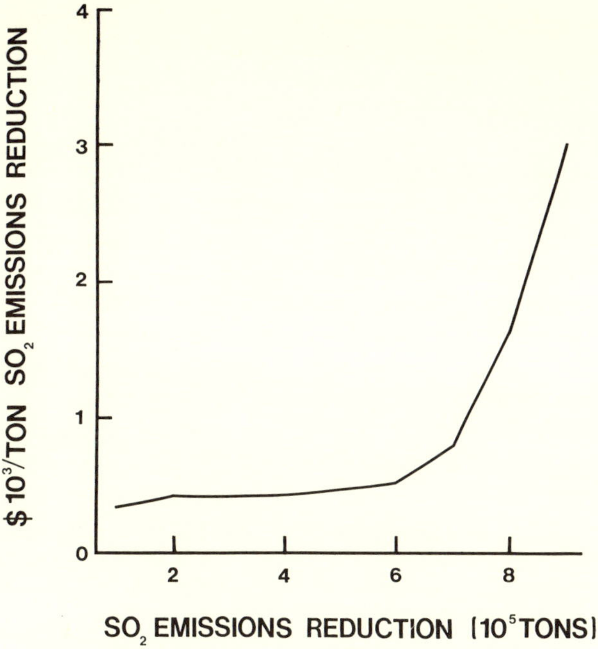

Figure 11.2 State Wide Marginal Costs of Sulfur Dioxide Emissions Control for Electric Utilities in Illinois (1983)

Table 11.1 - Allowable SO_2 Emissions (tons/year)

Utility	1.5 lb. excess emission reduction	50 percent proportional reduction	1.8 lb. systems cap
A	28,686	14,799	29,597
B	68,419	105,678	88,743
C	173,793	163,490	273,189
D	40,217	47,603	48,259
E	108,798	171,254	136,339
F	634	638	761
G	4,933	5,244	10,487
H	8,571	8,449	11,904
I	638	1,250	766
J	731	496	877
State total	435,520	518,978	600,922

ted by the regulatory control programs. From this point on, all discussion will address the 1.5 lbs. SO_2 excess emission reduction control program, but the procedure for the other two options is identical.

Table 11.2 ranks the pollution control options in order of marginal cost for each of the years between 1983 and 2000 for the 1.5 lbs. excess emissions strategy. The first control option, overcontrol on the scrubber on unit BE77-1, is the most efficient (it has the lowest marginal cost) in each of the 18 years. Similarly, the two other overcontrol options are at the top (ranking 2 and 3) in all subsequent years. The next three options, resulting from shifting to low-sulfur coal on the three units at utility A's plant site B, are consistently ranked as well.

The next option listed, a scrubber on unit CD67-1, is ranked seventh in terms of cost efficiency in the early years, drops to 13th by 1990, and returns to seventh by the conclusion of the study period. The reason for this drop will become apparent shortly. Similarly, the remaining control options through the 18th listed (low-sulfur coal at unit BD48-2) shift ranking on a year-to-year basis. After applying 18 control technologies in 1983, the statewide target for emissions (435,520 tons SO_2 per year) would be met, since emissions would be 433,260 tons/year. The emissions reduction associated with the 18th control option (3,293 tons SO_2 per year) is necessary, because the emission level would be above required levels if it were not employed.

In other years, however, if the final control option necessary to meet the target is a scrubber, then the level of control might greatly surpass the required reductions. For example, consider 1993. The portion of the marginal cost curve near the target level is outlined in Table 11.3. Control options through ranking 20 do not meet the allowable level, since emissions would be at 459,567 tons per year, above the 435,520 tons/year allowable level. The addition of the 21st control technology, however, drops the emissions total well below that required, so much that options 18, 19, and 20 are no longer really necessary. If we removed them, emissions levels would be 420,504 tons/year (that is, 406,905 + 4,734 + 4,633 + 4,232), a level still well below the allowable emission limitation. The ability to drop control options even though they have a lower marginal cost in a specific year is important, because we must examine the feasibility of using each option over time. To take the extreme case, the 18th option listed in Table 11.3 (a scrubber on unit GA63-5) appears only as a cost-efficient option for the one year—1993. In fact, utility G will add a new unit in 1995, and the only reason that the control option is cost-efficient in 1993 is that the unit is being operated at a relatively high level in those years to meet system load demands. After the new unit is brought on line, the capacity factor of unit GA63-3 will drop considerably. Since (a) the scrubber is not necessary before or after 1993, (b) the unit will be 30

Table 11.2 - Economically Efficient Control Technology for 1.5 lb. Excess Emissions Reduction Control Program

Utility unit	Control technology	Marginal cost ranking (Year 19__)																	
		83	84	85	86	87	88	89	90	91	92	93	94	95	96	97	98	99	2000
BE77-1	Overcontrol	1	1	1	1	1	1	1	1	1	1	1	1	1	1	1	1	1	1
AA76-1	Overcontrol	2	2	2	2	2	2	2	2	2	2	2	2	2	2	2	2	2	2
HA78-3	Overcontrol	3	3	3	3	3	3	3	3	3	3	3	3	3	3	3	3	3	3
AB68-2	Low-sulfur coal	4	4	4	4	4	4	4	4	4	4	4	4	4	4	4	4	4	4
AB72-3	Low-sulfur coal	5	5	5	5	5	5	5	5	5	5	5	5	5	5	5	5	5	5
AB60-1	Low-sulfur coal	6	6	6	6	6	6	6	6	6	6	6	6	6	6	6	6	6	6
CD67-1	Scrub	7	7	7	7	8	9	10	13	19	10	9	8	7	7	7	7	7	7
BA72-2	Low-sulfur coal	8	8	8	8	7	7	7	7	7	7	7	7	8	8	8	8	8	8
EA73-2	Scrub	9	11	10	13	13	13	14	9	13	11	11	16	10	11	12	11	14	10
BA65-1	Low-sulfur coal	10	10	9	9	9	8	8	8	8	8	8	9	9	9	9	9	9	9
EA70-1	Scrub	11	9	11	10	10	10	9	12	9	9	10	10	11	10	10	10	10	11
CD68-2	Scrub	12	12	14	15	17	17	18	20	17	17	17	15	15	14	11	12	11	13
BB58-2	Low-sulfur coal	13	13	12	11	11	11	11	10	11	12	12	11	12	12	13	13	12	12
BB50-1	Low-sulfur coal	14	14	13	12	12	12	12	11	12	13	13	12	13	13	14	14	13	-
BD60-3	Low-sulfur coal	15	16	16	14	14	14	13	14	14	14	14	13	14	15	15	15	15	14
BD48-1	Low-sulfur coal	16	17	17	16	15	15	15	15	15	15	15	14	16	16	-	-	-	-
EA75-3	Scrub	17	15	15	18	20	18	17	18	20	18	21	18	18	18	17	18	16	15
BD48-2	Low-sulfur coal	18	*	18	17	16	16	16	16	16	16	16	17	17	17	-	-	-	-
BC53-1	Low-sulfur coal					19			19	19		20					17		
BC54-2	Low-sulfur coal					18			17	18		19				16	16		
GA63-3	Scrub									18									

Note: Asterisk (*) represents marginal cost above market clearing price for that year, but control technology included in analysis for technical consistency.

Table 11.3 - Projected Statewide Aggregate Marginal Cost Curve Near 1.5 lb. Excess Emissions Allowable Level in 1993

Control option ranking	Unit	Control technology	Emission reduction (tons/yr)	Total state emissions (tons/year)	Marginal cost ($/ton SO_2 removed)
⋮	⋮	⋮	⋮	⋮	⋮
17	CD68-2	Scrub	64756	473166	523
18	GA63-3	Scrub	4734	468432	563
19	BC54-2	LSC	4633	463800	564
20	BC53-1	LSC	4232	459567	567
21	EA75-3	Scrub	52662	406905	610
⋮	⋮	⋮	⋮	⋮	⋮

Note: LSC = low-sulfur coal.

years old at that time, and (c) the load on the unit will drop shortly thereafter, it is apparent that the control option is not feasible from any viewpoint. Similarly, the use of low-sulfur coal only in selected years (1987, 1990, 1991, 1993, and 1998 in unit BC53-1) is not practical from the utility's viewpoint.

This suggests that while the year-by-year marginal cost data provide valuable information for selecting optimal control strategies, it must be tempered by "real world" considerations. Those control options in the shaded areas were dropped, even though they had slightly lower marginal costs, for the reasons discussed above. It should also be noticed that certain controls were employed even when not theoretically necessary, to provide technical consistency. For example, the emission limitation could have been met in 1984 without the 18th control option, use of low-sulfur coal in unit BD48-2. In all other thirteen years of operation this is required, however, so it was added in 1984 as well (and is indicated by an asterisk * in Table 11.2).

The first 18 control options were able to meet the statewide allowable emission limitation in all years. These thus form the economically efficient response to the 1.5 lbs. excess emission control program in the state.

RESPONSE OF THE INDIVIDUAL UTILITIES

Having determined an economically efficient control program response on a statewide basis to the 1.5 lbs. excess emission regulatory requirement, we now turn to the individual utilities to see how this optimum statewide response matches their allowable emissions limitations. This is outlined in Figure 11.3, which plots emissions profiles over time for each utility.

The top line in each graph represents a "status quo" emissions profile—that is, it assumes that existing emissions controls will remain in place, but no additional controls are added. An analysis of the profiles of the individual utilities is quite instructive and tells much about their future. Utilities A and F, for example, show a reasonably standard profile, with emissions increasing over time as the load grows. Utilities B and H show the same general profile, but show some decreases near the end of the century as older units are retired. Utility C has a number of new nuclear units coming on line through the mid- to late 1980s, and its coal-fired capacity factors will drop until the early 1990s. Growth will then lead to increased emissions until the very late 1990s, when new coal-fired capacity is added. It should be noted that this "dip" led to the shift in marginal costs rankings associated with controls on unit CD67-1 in Table 11.2.

Utility D, on the other hand, is a relatively older utility with only one major customer, the United States government. Its ownership is somewhat unusual in that it is owned by other utilities, and any future demand would probably be met by these parent firms rather than utility D. It is thus facing a decline in emissions over time after the early 1990s. Utility E will have a nuclear unit come on-line in the late 1980s, and coal-fired load will therefore drop at that time. It will continue to increase after that, however, because of load growth. As discussed earlier, Utility G will have a new unit come on-line in the mid-1990s. Just prior to that, emissions will increase as the older units are operated to meet the increasing load; after the new unit is added, they will drop substantially.

Utilities I and J are very small; Utility I expects its load to drop

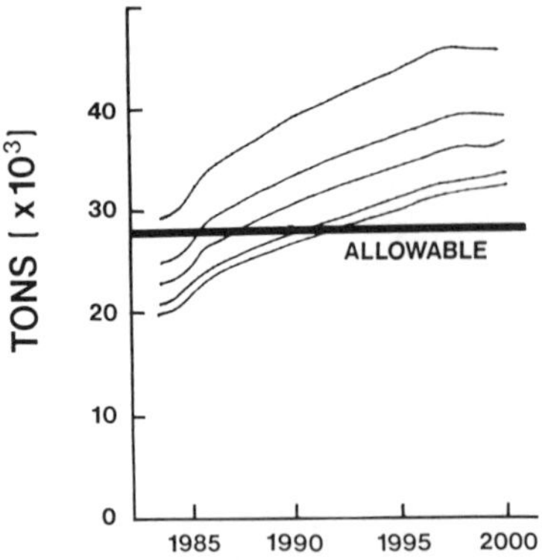

UTILITY A

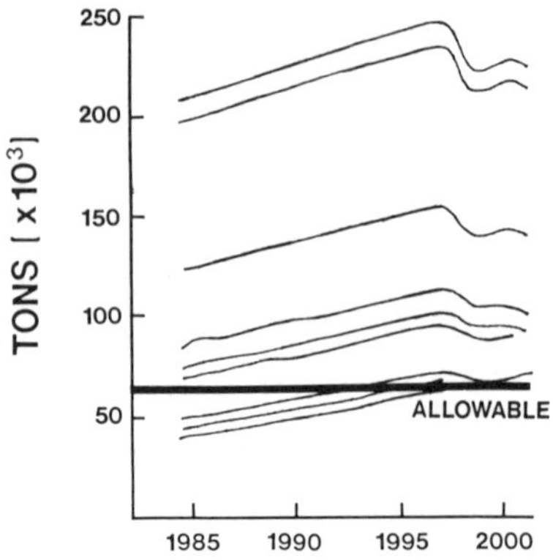

UTILITY B

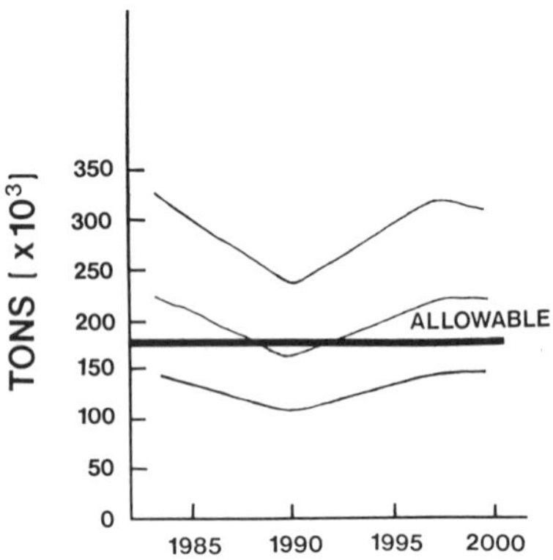

UTILITY C

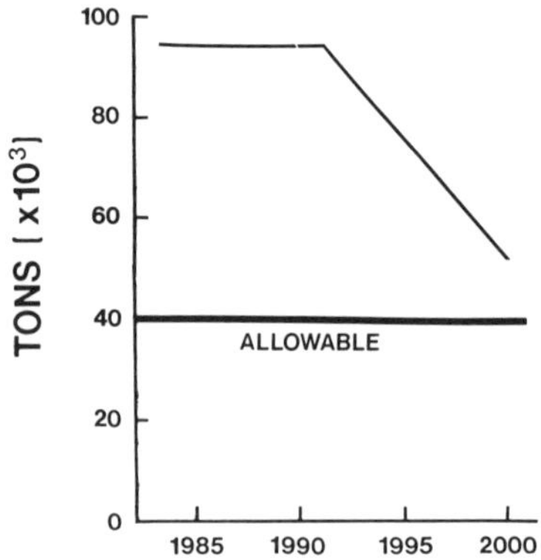

UTILITY D

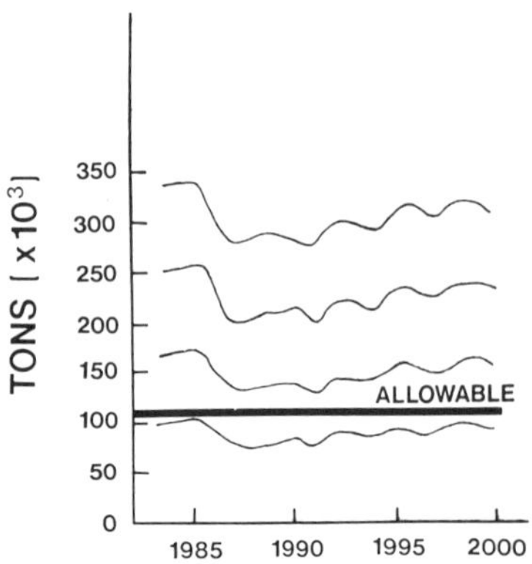

UTILITY E

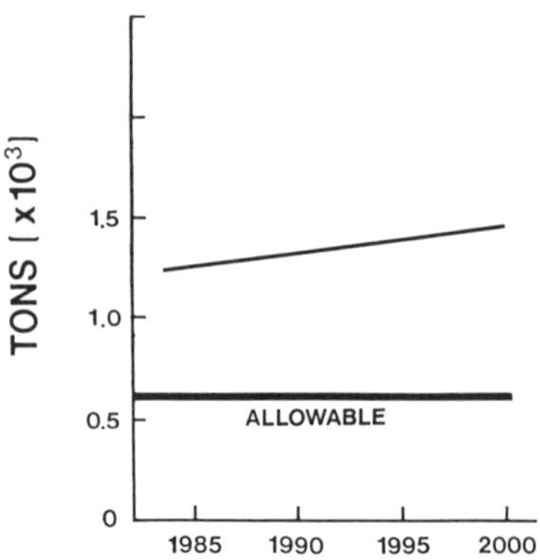

UTILITY F

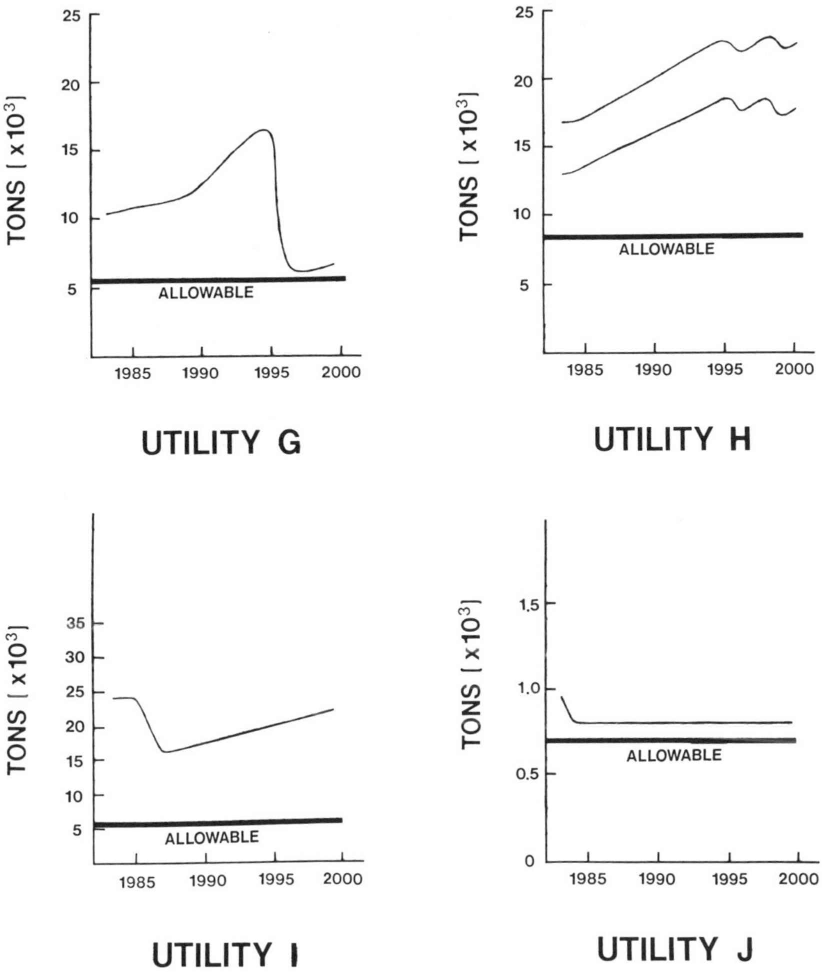

Figure 11.3 Economically Efficient Response of Individual Utilities to the 1.5 lb. Excess Emissions Control Program

when a new nuclear unit (owned by another utility in the general service area) comes on-line in the mid-eighties. Load should increase after that as growth continues. Utility J, on the other hand, expects to operate relatively constantly in the future: any growth would be handled by the major utility servicing the area. The slight decrease in emissions in 1984 is due to a turbine overhaul scheduled for that year, which should decrease the heat rate accordingly.

The curves below the top line in Figure 11.3 represent the emissions decreases associated with the 18 economically efficient control options discussed above. It can be seen that the 18 control options are distributed over only five utilities; the cost of controlling at the other five utilities is not efficient.

Utility A in Figure 11.3 shows a status quo profile and four control options. The control option closest to the status quo profile represents the one with the lowest marginal cost (that is, overcontrol in unit AA76-1). The second option (low-sulfur coal in unit AB68-2) has a higher marginal cost, and the remaining options are also ranked in order of marginal cost (the farther from the status quo line, the higher the marginal cost). All four are below the aggregate marginal cost necessary to meet the state allowable emission level, however. An interesting trend is apparent for Utility A in Figure 11.3. Note that the individual utility's allowable emissions could be met in 1983 by simply adding the first control option. By 1985, however, two control options must be placed in service, and by 1993 all four controls will be insufficient to meet the allowable emissions level for utility A. In the market sense, Utility A shifted from being a potential seller of SO_2 ERCs to a buyer. We shall return to this point shortly.

THE MARKET RESPONSE

While Utility A could meet the regulatory requirements of a 1.5 lbs. excess emissions limitation in 1983 with one control option (with considerable room to spare), the cost of implementing the other three options is still below the market price (the aggregate marginal cost of control in the state at the allowable level). Thus, it would be worthwhile for one of the five utilities who do not have economically efficient control options to pay utility A to achieve the reductions; utility A would gain (market price less control costs), and the buying utility would also save (inefficient control costs less market price), for every ton of SO_2 reduction. The likely response of the individual utilities is outlined in Table 11.4. This table shows that utilities A and B both shift from being sellers to buyers. Utilities C and E dominate the supply of SO_2 ERCs, while utility D is the major buyer.

Table 11.4 - Market Response of Individual Utilities to 1.5 lb. $SO_2/10^6$ "Excess" Emission Reduction

Utilities	BUY											SELL									
	SO_2 ERCS (tons/year)																				
	A	B	C	D	E	F	G	H	I	J	Total	A	B	C	D	E	F	G	H	I	J	Total
1983	-	-	-	55149	-	641	5554	4671	1861	260	68136	8359	26500	28924	-	6613	-	-	-	-	-	70396
1984	-	-	-	55149	-	655	5749	4690	1861	105	68209	7542	25538	35083	-	5896	-	-	-	-	-	74059
1985	-	-	-	55149	-	667	5945	5271	1861	105	68998	5404	24574	40768	-	1788	-	-	-	-	-	72534
1986	-	-	-	55149	-	679	6140	5854	1413	105	69340	4396	23494	46453	-	18684	-	-	-	-	-	93027
1987	-	-	-	55149	-	694	6333	6242	1040	105	69563	3499	20974	52612	-	30335	-	-	-	-	-	107420
1988	-	-	-	55149	-	706	6529	6721	1081	105	70291	2528	19425	58297	-	27944	-	-	-	-	-	108194
1989	-	-	-	55149	-	720	6803	7266	1126	105	71169	1767	17742	63982	-	27148	-	-	-	-	-	110639
1990	-	-	-	55149	-	734	7683	7598	1171	105	72440	1232	15946	70141	-	22211	-	-	-	-	-	109530
1991	-	-	-	55149	-	746	8563	8019	1215	105	73797	565	13973	64929	-	33234	-	-	-	-	-	112701
1992	133	-	-	50284	-	761	9590	8576	1260	105	70709	-	11973	60192	-	20706	-	-	-	-	-	92871
1993	600	-	-	45420	-	775	10620	9194	1309	105	68023	-	9909	55454	-	17677	-	-	-	-	-	83040
1994	1220	-	-	40556	-	789	11650	9753	1357	105	65430	-	7879	50243	-	23853	-	-	-	-	-	81975
1995	1911	-	-	35691	-	803	11702	10214	1406	105	61832	-	6050	45032	-	13595	-	-	-	-	-	64677
1996	2575	-	-	30827	-	818	4173	9087	1458	105	49043	-	4303	40294	-	14153	-	-	-	-	-	58750
1997	3118	-	-	25962	-	832	1291	9589	1510	105	42407	-	1233	35629	-	19457	-	-	-	-	-	56319
1998	3573	-	-	21098	-	846	1394	10224	1562	105	38802	-	1636	31579	-	8298	-	-	-	-	-	41513
1999	3878	-	-	16233	-	860	1521	8816	1618	105	33031	-	326	29106	-	8627	-	-	-	-	-	38059
2000	4300	468	-	11821	-	877	1623	9230	1674	105	30098	-	-	26649	-	12464	-	-	-	-	-	39113

The data are shown graphically in Figure 11.4, with potential "sellers" above the X axis, and potential "buyers" below it. Only four utilities are above the X axis, the potential sellers A, B, C, and E. If their profiles are examined separately, as in Figure 11.5, it can be seen that both A and B require the emissions reductions from their economically efficient control technologies in the later years of the study period (that is, both ultimately become buyers). It is apparent, then, that these two firms will not be willing to sell the ERCs associated with these control technologies, since they must be assured that these ERCs will be available in future years. This can be done in only two ways: (a) a sales agreement with the stipulation that the ERC will be sold back at a specified later period (termed a "contingent sale"), or (b) a leasing arrangement. For the remainder of this section, the analysis will discuss results in terms of the leasing arrangement, although the contingent sale would also be applicable.

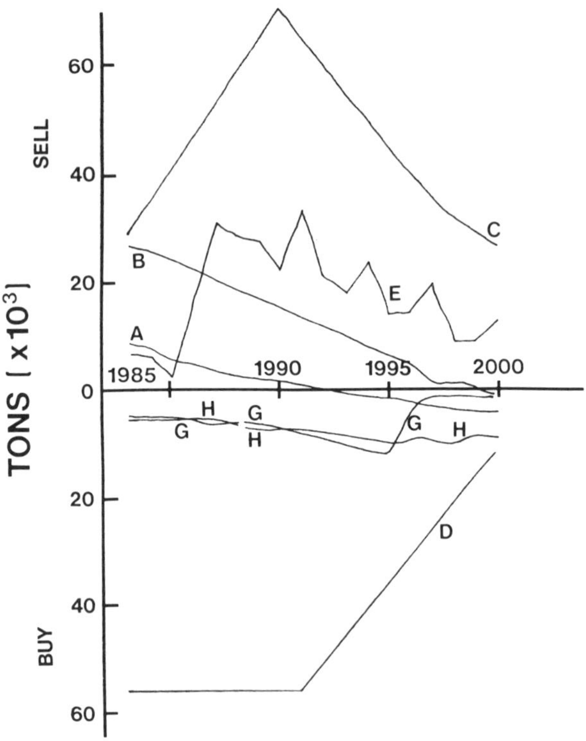

Figure 11.4 Potential SO_2 ERCs in the 1.5 lb. Excess Emissions Reduction Control Program

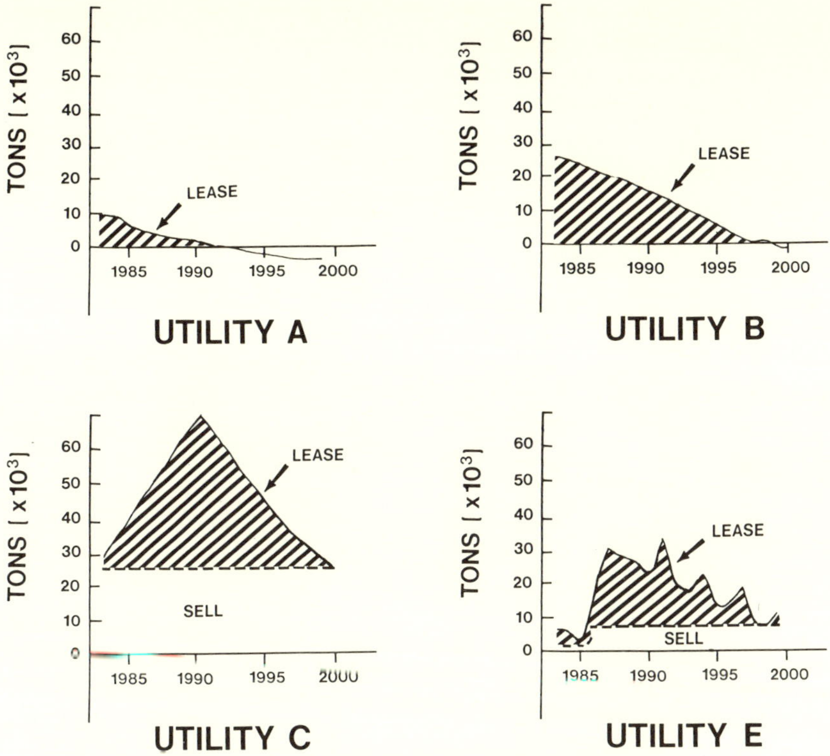

Figure 11.5 ERCs Available for Sale or Lease Under the 1.5 lb. per MM BTU Excess Emissions Reduction Control Program

Utilities C and E are in a similar position, but they could offer some ERCs for sale. It is important to note, however, that these utilities will also be willing to sell only those ERCs not required in the future years. The distribution between those available for sale and those for lease is clearly shown in Figure 11.5. The maximum quantity of ERCs that would be placed on the market for sale can thus be determined, and these are listed in Table 11.5.

It can be seen from Table 11.4, however, that the statewide demand for SO_2 ERCs will exceed the amount for sale in 16 of the 18 years in the study period (that is, 1983–1998). This implies that emissions trading sufficient to meet the control requirement in an economically efficient manner will not occur in these years unless the desire for utilities to maintain control of their ERCs in future years is taken into account. Thus, a contingent sale or lease is a necessary component of the market in achieving cost efficiency.

Table 11.5 - Maximum ERCs Available for Sale Under 1.5 lb. Excess Emission Regulatory Control Program

Utility	Potential SO_2 ERCs (tons/year)
A	-
B	-
C	26649
D	-
E	8298
F	-
G	-
H	-
I	-
J	-
State total	34947

While Table 11.5 indicates that 34,947 tons of SO_2 ERC could be available for sale, the actual figure is likely to be considerably less. Note that the general trend of the curves for Utilities C and E in Figure 11.5 are downward sloping in the final half of the study period. If the period were extended beyond the year 2000, the amount available for sale by these two utilities would probably drop. Similarly, a previous interview with utility C indicated a reluctance to even consider selling ERCs. It is quite conceivable that very few (or even no) SO_2 ERCs will be available for sale, and the entire market may well require leasing arrangements.

While the discussion thus far has focused on the supply of ERCs, it is also important to address demand. Between six and eight of the utilities are "buyers," depending upon the specific year. Three of these (F, I and J) are quite small, accounting for only a few percent of the demand in any given year. The market is dominated by utility D, particularly in the early years of the study period. These results are particularly promising, since utility D exhibits exactly the type of demand profile conducive to leasing. The utility has a high demand in early years, when other utilities (A and B) have ERCs available. In later years, however, when these utilities need their ERCs and all of the "selling" utilities have downward slopes, utility D's demand drops considerably. The utility should therefore be willing to enter into a leasing arrangement for the ERCs.

A similar pattern is found for utility G, although the emission quantities involved are much smaller. The utility requires a considerable amount of ERCs in the early 1990s, as it uses its older units to meet growing load demands; when a new unit is brought on line replacing this capacity, however, the demand for ERCs drops dramatically. This utility should also be willing to lease, at least to the level of emissions eliminated when the new unit comes on line. The common thread of the two utilities willing to lease is the retirement of older, higher emissions capacity. This is discussed in a later section.

OTHER ACID DEPOSITION CONTROL PROGRAMS

The results thus far have strictly been applicable to the 1.5 lbs. excess emissions regulatory control program. An identical analysis was conducted for the 50 percent proportional reduction and 1.8 lbs. systems cap, which are slightly less stringent. The results of these analyses are outlined below.

The 50 Percent Proportional Reduction Control Program

The economically efficient control technology for a 50 percent proportional reduction is outlined in Table 11.6. It is obviously quite similar to Table 11.2, with the exception that three fewer technologies are required. It should also be noted that the scrubber on CD68-2 is included for a five-year period in the late 1980s and early 1990s, even though its marginal cost is higher than other control options (for example, low-sulfur coal on units BD48-1 and -2) in order to obtain consistency over time. As noted earlier, this corresponds to the period of time when utility C has a "dip" in its coal-fired capacity factors because of new nuclear units coming on line.

The effects of these technologies on the individual utilities, along with allowable emission levels under this control program, are outlined in Figure 11.6. As noted earlier, this strategy hurts utilities who have already taken control measures, and assists those who haven't. For example, in the previous 1.5 lbs. excess emission control program, both utility A and utility B began the study period as sellers and ended as buyers. In this strategy, however, utility A, which has a 1.4 lbs. per MM BTU system average in the base year, becomes a buyer in all years. Utility B, with a much higher 4.3 lbs. per MM BTU system average in the base year, becomes a seller in all years.

The amounts bought and sold are outlined in Table 11.7. Once again D is the major buyer, and C and E are sellers. The market response is graphically shown in Figure 11.7. Note that the potential ERCs avail-

Table 11.6 - Economically Efficient Control Technology for 50% Proportional Reduction Control Program

Utility unit	Control technology	Marginal cost ranking (Year 19__)																	
		83	84	85	86	87	88	89	90	91	92	93	94	95	96	97	98	99	2000
BE77-1	Overcontrol	1	1	1	1	1	1	1	1	1	1	1	1	1	1	1	1	1	1
AA76-1	Overcontrol	2	2	2	2	2	2	2	2	2	2	2	2	2	2	2	2	2	2
HA78-3	Overcontrol	3	3	3	3	3	3	3	3	3	3	3	3	3	3	3	3	3	3
AB68-2	Low-sulfur coal	4	4	4	4	4	4	4	4	4	4	4	4	4	4	4	4	4	4
AB72-3	Low-sulfur coal	5	5	5	5	5	5	5	5	5	5	5	5	5	5	5	5	5	5
AB60-1	Low-sulfur coal	6	6	6	6	6	6	6	6	6	6	6	6	6	6	6	6	6	6
CD67-1	Scrub	7	7	7	7	8	9	10	13	10	10	9	8	7	7	7	7	7	7
BA72-2	Low-sulfur coal	8	8	8	8	7	7	7	7	7	7	7	7	8	8	8	8	8	8
EA73-2	Scrub	9	11	10	13	13	13	14	9	13	11	11	16	10	11	12	11	14	10
BA65-1	Low-sulfur coal	10	10	9	9	9	8	8	8	8	8	8	9	9	9	9	9	9	9
EA70-1	Scrub	11	9	11	10	10	10	9	12	9	9	10	10	11	10	10	10	10	11
CD68-2	Scrub	12	12	14	15	*	*	*	*	*	17	17	15	15	14	11	12	11	13
BB58-2	Low-sulfur coal	13	13	12	11	11	11	11	10	11	12	12	11	12	12	*	13	12	12
BB50-1	Low-sulfur coal	14	14	13	12	12	12	12	11	12	13	13	12	13	13	*	*	13	-
BD60-3	Low-sulfur coal	15	*	*	14	14	14	13	14	14	14	14	13	14	15	*	*	15	*
BD48-1	Low-sulfur coal					15	15	15	15		15	15	14						
EA75-3	Scrub		15	15															
BD48-2	Low-sulfur coal						16				16	16							

Note: Asterisk (*) represents marginal cost above market clearing price for that year, but control technology included in analysis for technical consistency.

able from C and E have dropped, while B has moved into a position as a supplier. Utility A now becomes a major factor in the demand. Once again, it is apparent that leasing arrangements (or contingency sales) are required.

The maximum quantity of SO_2 ERCs available for sale can be similarly totaled, and these are outlined in Table 11.8. Note that utility B is now the principal supplier. The quantity of ERCs available for sale is higher than the previous control program, but is still not sufficient to meet the demand in 14 of the 18 years (1983 to 1996). Once again, it is apparent that leasing arrangements (or contingency sales) are required.

The 1.8 lbs. System Cap Control Program

Comparable results were obtained for the 1.8 lbs. systems cap. The economically efficient control technology is shown in Table 11.9, and the major change is that two more control technologies can be dropped (including the CD68-2 scrubber discussed earlier). By dropping this scrubber, however, the state would not quite achieve compliance in the first three years (the levels achieved would be 607,696 tons SO_2 in 1983, 606,331 tons SO_2 in 1984 and 607,340 tons SO_2 in 1985—approximately 98.5 percent of the required emission reduction). Given the lead times for scrubber construction, the probable compliance times in proposed acid deposition legislation, and the fact that this scrubber would not be necessary in the remaining 15 years, non-compliance was deemed a more realistic prospect than the alternative.

The result of this type of regulatory program is particularly beneficial for utility C, even beyond the elimination of the scrubber on CD68-2. Figure 11.8 shows the impact on individual utilities, and it can be seen that no control technology at all would be required for utility C for the years 1988 to 1993. The shift to noncoal-fired generation could thus be an important factor for this firm.

Table 11.10 outlines potential ERC transactions that would occur under this control program. Once again, utilities A and B are sellers initially, but become buyers near the end of the study period. Once again, utility D is the major buyer of ERCs. The principal result, however, shown even more clearly in Figure 11.9, is the complete domination by utility C as the seller of SO_2 ERCs in the state. If the maximum potential ERCs available are determined, as in Table 11.11, it becomes clear that utility C is virtually a monopolist in SO_2 ERCs.

Several caveats should be made at this time. First, utility C was the firm mentioned earlier that did not provide production cost modeling data, and the estimation technique for both total system growth and its allocation to individual generating stations may contain unknown errors. Second, the pollution control agency might balk at giving ERCs

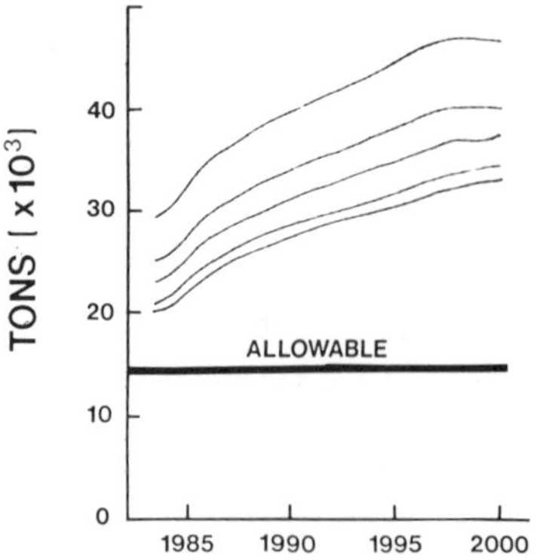

UTILITY A

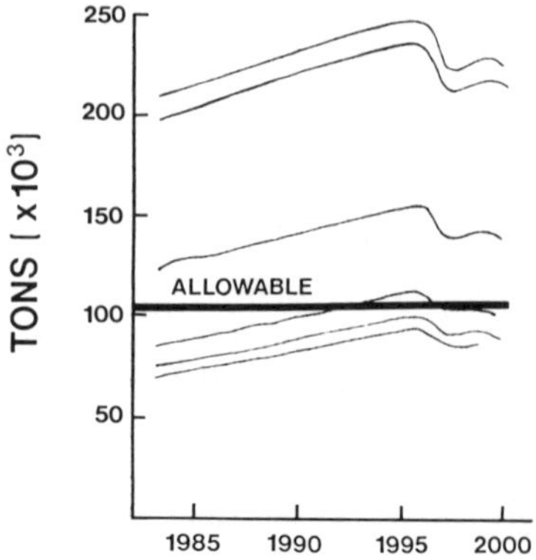

UTILITY B

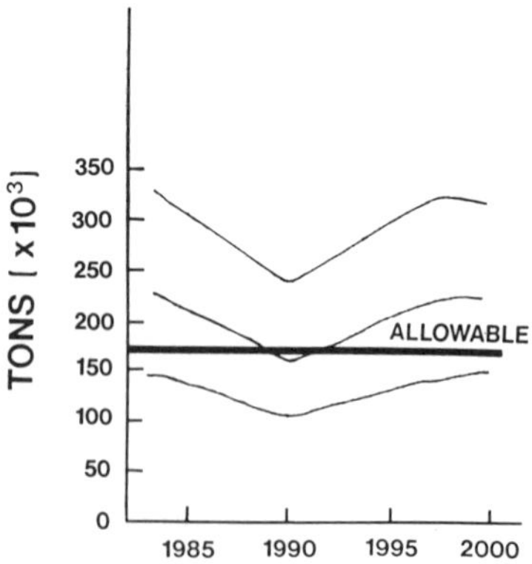

UTILITY C

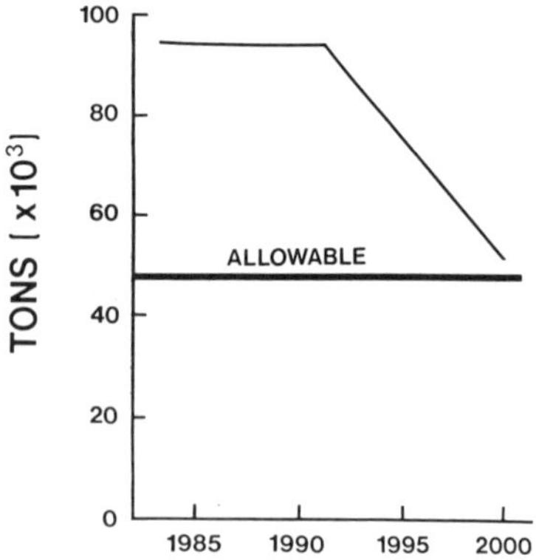

UTILITY D

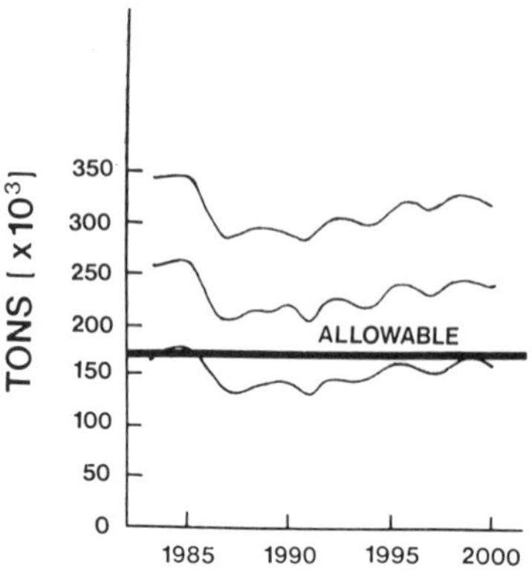

UTILITY E

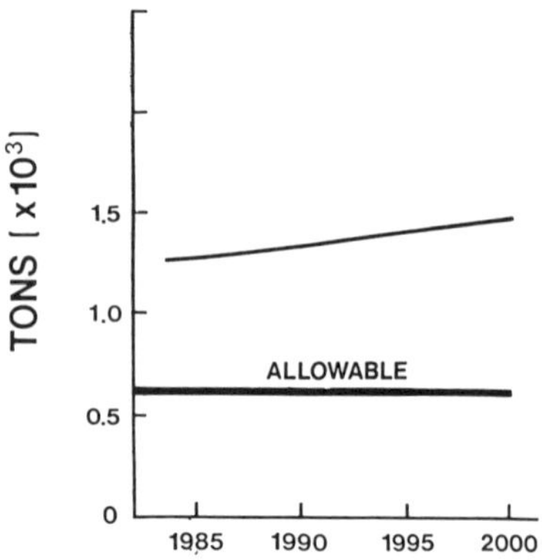

UTILITY F

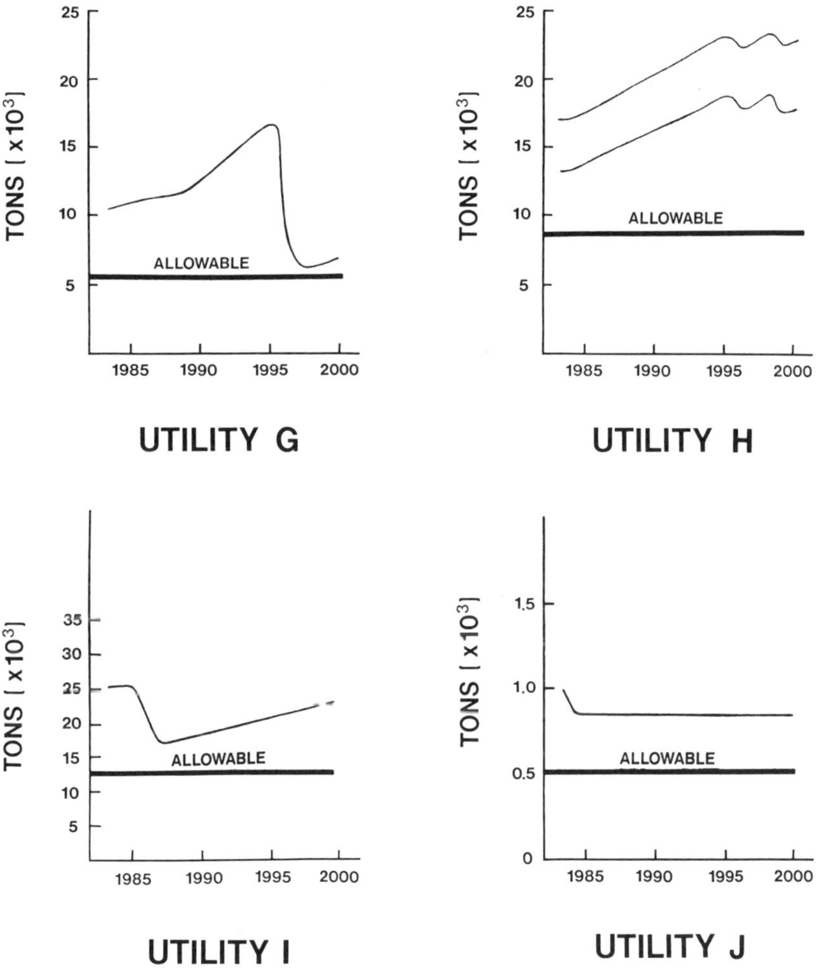

Figure 11.6 Economically Efficient Response of Individual Utilities to the 50% Proportional Reduction Control Program

Table 11.7 - Market Response of Individual Utilities to 50% Proportional Emissions Reduction

SO_2 ERCS (tons/year)

Utilities	BUY										SELL										
	A	B	C	D	E	F	G	H	I	J	Total	A	B	C	D	E	F	G	H	I	J	Total
1983	5528	-	-	47683	-	637	5243	4793	1249	495	65628	-	55945	18521	-	2304	-	-	-	-	-	76770
1984	6345	-		47683	2053	651	5438	4812	1249	340	68571	-	54394	24680	-	-	-	-	-	-	-	79074
1985	8483	-	-	47683	7654	663	5634	5393	1249	340	77099	-	53289	30365	-	-	-	-	-	-	-	83654
1986	9491	-	-	47683	-	675	5829	5976	801	340	70795	-	52063	36050	-	19679	-	-	-	-	-	107792
1987	10388	-	-	47683	-	690	6022	6364	428	340	71915	-	49900	42209	-	41319	-	-	-	-	-	133428
1988	11359	-	-	47683	-	702	6218	6843	469	340	73614	-	48253	47894	-	33640	-	-	-	-	-	129787
1989	12120	-	-	47683	-	716	6492	7388	514	340	75253	-	46414	53579	-	31469	-	-	-	-	-	131462
1990	12655	-	-	47683	-	730	7372	7720	559	340	77059	-	44475	59738	-	28806	-	-	-	-	-	133019
1991	13322	-	-	47683	-	742	8252	8141	603	340	79083	-	42463	54526	-	42851	-	-	-	-	-	139840
1992	14020	-	-	42818	-	757	9279	8698	648	340	76560	-	40359	49789	-	24248	-	-	-	-	-	114396
1993	14487	-	-	37954	-	771	10309	9316	697	340	73874	-	38317	45051	-	27471	-	-	-	-	-	110839
1994	15107	-	-	33090	-	785	11339	9875	745	340	71281	-	36346	39840	-	26846	-	-	-	-	-	103032
1995	15798	-	-	28225	-	799	11391	10336	794	340	67683	-	34562	34629	-	12559	-	-	-	-	-	81750
1996	16462	-	-	23361	-	814	3862	9209	846	340	54894	-	32844	29891	-	11610	-	-	-	-	-	74345
1997	17005	-	-	18496	-	828	980	9711	898	340	48258	-	38492	25226	-	23465	-	-	-	-	-	87183
1998	17460	-	-	13632	-	842	1083	10346	950	340	44653	-	38895	21176	-	12963	-	-	-	-	-	73034
1999	17765	-	-	8767	-	856	1210	8938	1006	340	38882	-	37585	18703	-	2535	-	-	-	-	-	58823
2000	18187	-	-	4355	-	873	1312	9352	1062	340	35481	-	36791	16246	-	12292	-	-	-	-	-	65329

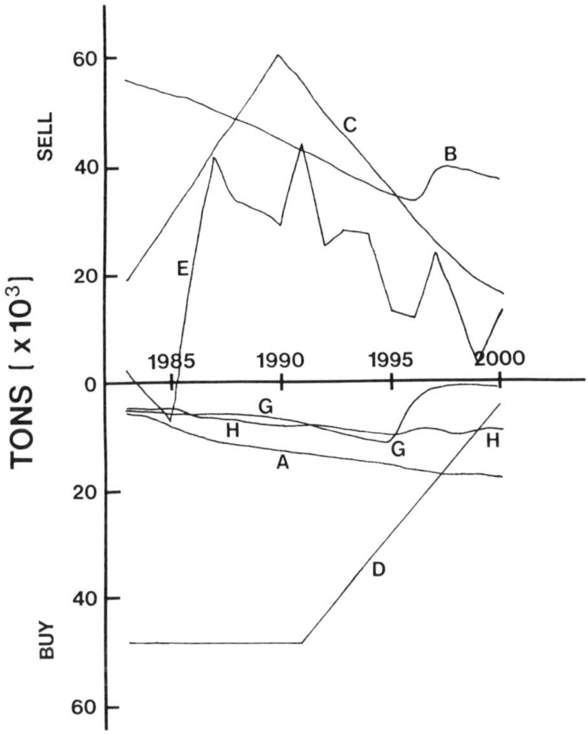

Figure 11.7 Potential SO_2 ERCs in the 50% Proportional Reduction Control Program

for the shift in generation at utility C (even though the utility might willingly take a permit constraint), taking the position that the reduction would have occurred in any event. Utility C would still have the ERCs associated with its scrubber operation, and the resulting profile (the dashed line in Figure 11.9) would not change the results substantively. Total ERCs (available for both sale and lease) would still be able to meet demand.

Third, the 41 tons per year SO_2 ERCs available from utility J are also an anomaly, that would probably not be credited in an actual control program. As noted above, utility J did not add control technology or shift its fuel to obtain the lower emissions level; instead, it upgraded its turbine and subsequently reduced its heat rate. The ERCs have been included in the analysis to maintain consistency, but the fact that the utility was still operating at 2.0 lbs. per MM BTU after this 1.8 lbs.

Table 11.8 - Maximum ERCs Available for Sale Under a 50% Proportional Reduction Regulatory Control Program

Utility	Potential SO_2 ERCs (tons/year)
A	-
B	32844
C	16246
D	-
E	2535
F	-
G	-
H	-
I	-
J	-
State total	51625

system cap requirement was instituted would probably cause some concern at the regulatory agency. In any event, the utility would be a very small buyer (or seller) of ERCs, and would have a negligible impact on the market.

Finally, as noted earlier, utility C has indicated that it is not particularly interested in emissions trading and turned down previous opportunities when approached. It may be in a position, due to its extensive investment in nuclear power, to meet emissions limitations at plant site D through capacity constraints; the firm also indicated that it might seriously consider retiring the plant instead of installing scrubbers, if the compliance deadline were sufficiently long. It is interesting to note that this option could result in even greater amounts of SO_2 ERCs. The firm's monopolist position might cause it to reconsider its position on emissions trading, but it is unlikely that the utility could take advantage of its monopoly power in such a regulated environment.

With these factors in mind, it can nevertheless be seen that, once again, the total supply of ERCs available for sale will not be sufficient to meet demand in 13 of the 18 years (1983 through 1986 and 1988 through 1996). Leasing or contingent sales will be required for this control program as well.

Table 11.9 - Economically Efficient Control Technology for 1.8 lb. Systems Cap Control Program

Utility unit	Control technology	Marginal cost ranking (Year 19__) 83	84	85	86	87	88	89	90	91	92	93	94	95	96	97	98	99	2000
BE77-1	Overcontrol	1	1	1	1	1	1	1	1	1	1	1	1	1	1	1	1	1	1
AA76-1	Overcontrol	2	2	2	2	2	2	2	2	2	2	2	2	2	2	2	2	2	2
HA78-3	Overcontrol	3	3	3	3	3	3	3	3	3	3	3	3	3	3	3	3	3	3
AB68-2	Low-sulfur coal	4	4	4	4	4	4	4	4	4	4	4	4	4	4	4	4	4	4
AB72-3	Low-sulfur coal	5	5	5	5	5	5	5	5	5	5	5	5	5	5	5	5	5	5
AB60-1	Low-sulfur coal	6	6	6	6	6	6	6	6	6	6	6	6	6	6	6	6	6	6
CD67-1	Scrub	7	7	7	7	8	9	10	13	10	10	9	8	7	7	7	7	7	7
BA72-2	Low-sulfur coal	8	8	8	8	7	7	7	7	7	7	7	7	8	8	8	8	8	8
EA73-2	Scrub	9	11	10	13	13	13	*	9	13	11	11	*	10	11	*	11	*	10
BA65-1	Low-sulfur coal	10	10	9	9	9	8	8	8	8	8	8	9	9	9	9	9	9	9
EA70-1	Scrub	11	9	11	10	10	10	9	12	9	9	10	10	11	10	10	10	10	11
CD68-2	Scrub	12	12	14									15			11	12	11	
BB58-2	Low-sulfur coal	*	*	12	11	11	11	11	10	11	*	*	11	12	12	*	*	12	*
BB50-1	Low-sulfur coal	*	*	13	12	12	12	12	11	12	*	*	12	13	13	*	*	*	-
BD60-3	Low-sulfur coal							13					13						
BD48-1	Low-sulfur coal												14						

Note: Asterisk (*) represents marginal cost above market clearing price for that year, but control technology included in analysis for technical consistency.

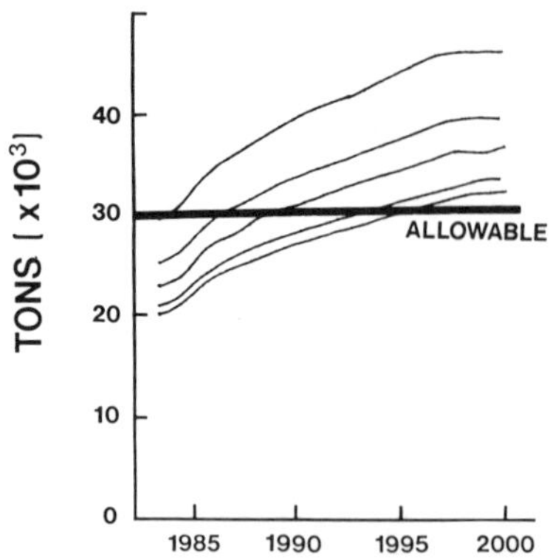

UTILITY A

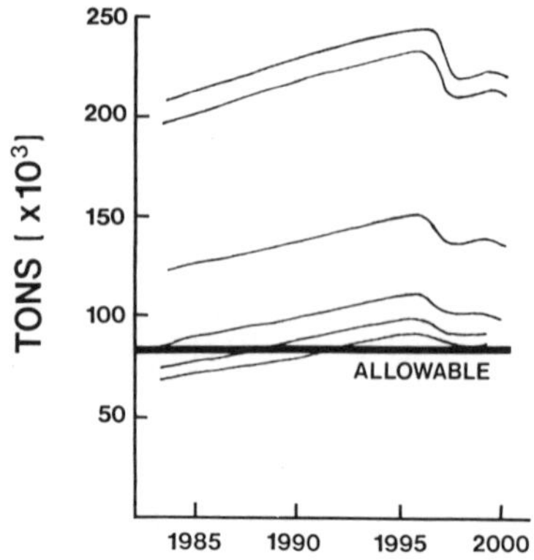

UTILITY B

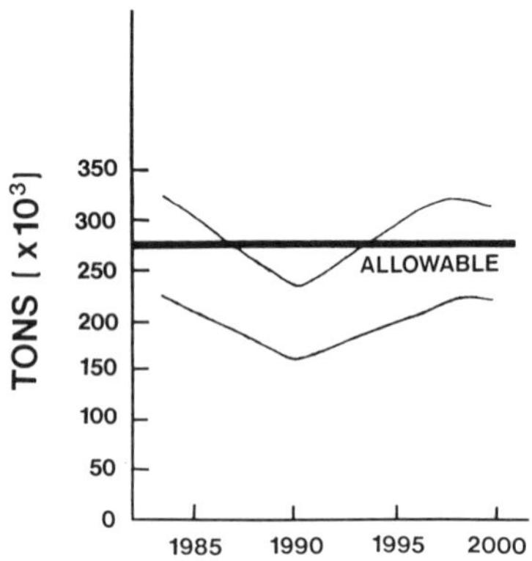

UTILITY C

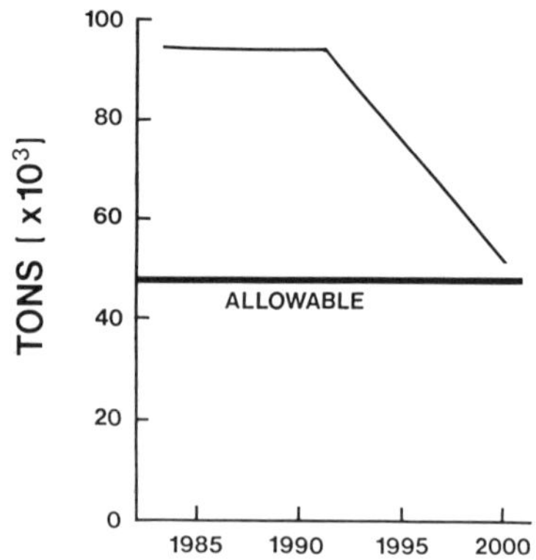

UTILITY D

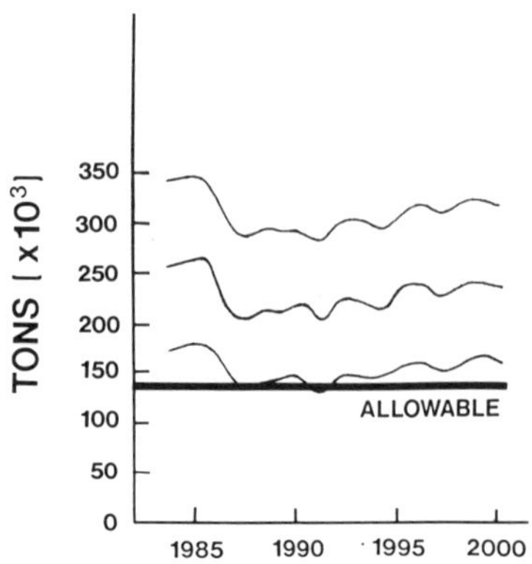

UTILITY E

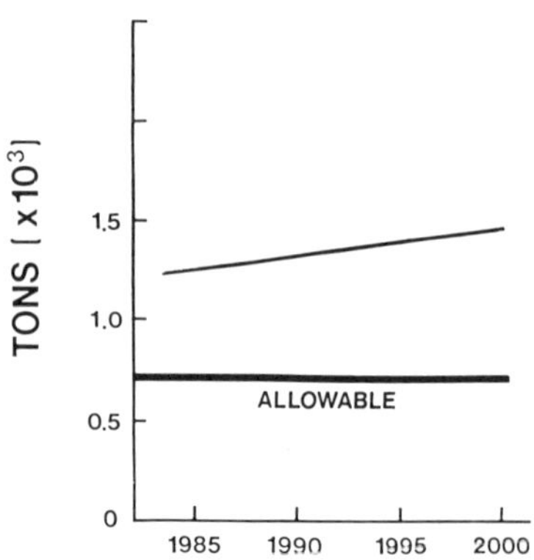

UTILITY F

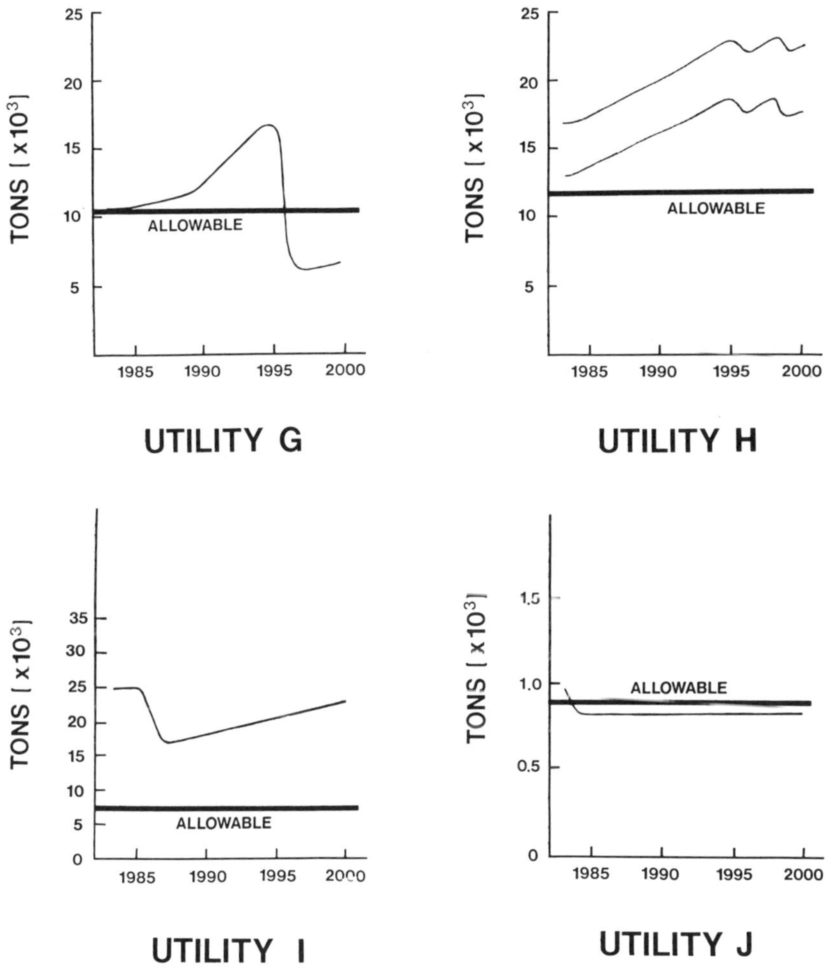

Figure 11.8 Economically Efficient Response of Individual Utilities to the 1.8 lb. Systems Cap Control Program

Table 11.10 - Market Response of Individual Utilities to 1.8 lb. $SO_2/10^6$ BTU System Cap

SO_2 ERCS (tons/year)

Utilities	BUY											SELL									
	A	B	C	D	E	F	G	H	I	J	Total	A	B	C	D	E	F	G	H	I	J	Total
1983	-	-	-	47107	32611	514	-	1338	1733	114	83417	9270	18413	48959	-	-	-	-	-	-	-	76642
1984	-	-	-	47107	36968	528	195	1357	1733	-	87888	8453	15500	58485	-	-	-	-	-	-	41	82479
1985	-	-	-	47107	42569	540	391	1938	1733	-	94278	6315	14225	67279	-	-	-	-	-	-	41	87860
1986	-	-	-	47107	15236	552	586	2521	1285	-	67287	5307	12849	76072	-	-	-	-	-	-	41	94269
1987	-	-	-	47107	-	567	779	2909	912	-	52274	4410	10400	85598	-	6404	-	-	-	-	41	106853
1988	-	-	-	47107	1275	579	975	3388	953	-	54277	3439	8598	94391	-	-	-	-	-	-	41	106469
1989	-	-	-	47107	3446	593	1249	3933	998	-	57326	2678	6617	103185	-	-	-	-	-	-	41	112521
1990	-	-	-	47107	6109	607	2129	4265	1043	-	61260	2143	4550	112711	-	-	-	-	-	-	41	119445
1991	-	-	-	47107	-	619	3009	4686	1087	-	56508	1476	2415	104650	-	7936	-	-	-	-	41	116518
1992	-	-	-	42242	10667	634	4036	5243	1132	-	63954	778	192	97322	-	-	-	-	-	-	41	98333
1993	-	1935	-	37378	7444	648	5066	5861	1181	-	59513	311	-	89995	-	-	-	-	-	-	41	90347
1994	309	3990	-	32514	8069	662	6096	6420	1229	-	59289	-	-	81934	-	-	-	-	-	-	41	81975
1995	1000	5827	-	27649	22356	676	6148	6881	1278	-	71815	-	-	73874	-	-	-	-	-	-	41	73915
1996	1664	7579	-	22785	23305	691	-	5754	1330	-	63108	-	-	66546	-	-	-	1381	-	-	41	67968
1997	2207	330	-	17920	11450	705	-	6256	1382	-	40250	-	-	58535	-	-	-	4263	-	-	41	62839
1998	2662	-	-	13056	21952	719	-	6891	1434	-	46714	-	235	54103	-	-	-	4160	-	-	41	58539
1999	2967	1302	-	8191	32380	733	-	5483	1490	-	52546	-	-	54006	-	-	-	4033	-	-	41	58080
2000	3389	2472	-	3779	22623	750	-	5897	1546	-	40456	-	-	53923	-	-	-	3931	-	-	41	57895

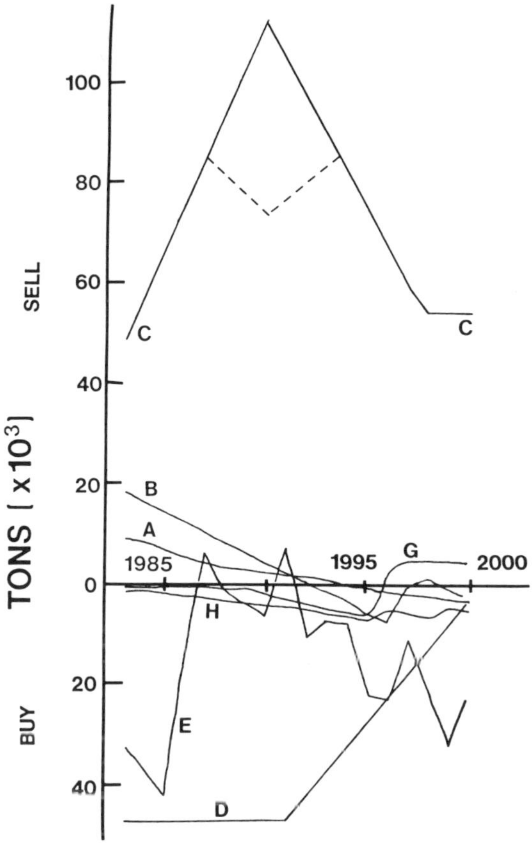

Figure 11.9 Potential SO_2 ERCs in the 1.8 lb. Systems Cap Control Program

CONTROL COSTS

The costs and savings associated with each control program, assuming emissions trading is employed, are outlined in Table 11.12. As one would expect, the costs decrease as the stringency of the program decreases. It is difficult to compare these figures directly with other cost estimates, given differences such as the base years of the analyses, the various aggregations and data reporting approaches (for example, total revenue requirements or percent rate increases) and even the ranges of the data presented (one estimate shows Illinois control costs ranging from \$30 to \$300 million/year).[1] Nonetheless, our annualized total control costs appear to be higher than those estimated by ICF, Inc. Their analysis of S-3041, similar to the 1.5 lbs. excess emission

Table 11.11 - Maximum ERCs Available For Sale Under a 1.8 lb. $SO_2/10^6$ BTU Systems Cap

Utility	Potential SO_2 ERCs (tons/year)
A	-
B	-
C	53923
D	-
E	-
F	-
G	-
H	-
I	-
J	41
State total	53964

reduction outlined previously, showed annualized costs of $69 million per year in Illinois for 1995 (this and all subsequent costs are based on 1982 dollars), with a statewide marginal cost of $480 per ton of SO_2 removed.[2] Our analysis outlined above has annualized costs of approximately $250 million per year at the same period, with a marginal cost of $522 per ton of SO_2 removed. Some of this is due to differences in assumed reductions required (471,000 tons SO_2 for ICF, versus 582,000 tons SO_2 in this analysis), our reliance on utility cost data (which is

Table 11.12 - Annualized Control Costs and Savings in Illinois Through Emissions Trading

Control program	Annualized costs (10^6 dollars)[a]	Annualized savings (10^6 dollars)[a]
1.5 lb. excess emissions	244 - 251	54 - 78
50% proportional reduction	205 - 213	58 - 82
1.8 lb. systems cap	160 - 168	75 - 112

[a]1982 dollars

almost certainly higher), and differences in the sophistication of the cost models.

Since the firms with the lowest marginal cost of control implement the sulfur dioxide control options under the emissions trading scenarios, costs are lower than if each individual utility had to meet its allowable level independently. These costs vary from year to year, but estimated savings are also listed in Table 11.12. It should be noted that these cost differences are only approximations because of factors such as the noncompliance issue in the first three years of the study period for the 1.8 lb. system cap and similar problems that occur for specific utilities under some nontrading scenarios.

ICF's cost for an emissions cap, which allows no emissions trading, is \$189 million per year; this compares with a similar cost of \$321 million for this analysis. While these results are somewhat closer, the resulting differences from those discussed earlier suggest that intrastate emission trading offers savings of approximately \$120 million per year according to ICF, a value even greater than those outlined in Table 11.12.

In addition to the annualized costs and savings, it is important to remember that a wealth transfer occurs as buying utilities pay selling (or leasing) utilities for ERCs. The statewide total of these transfer payments are summarized in Table 11.13. These payments were estimated by multiplying the marginal cost of control achieving the allowable level times the demand for ERCs in each year under each scenario. They thus do not include the effect of the differences between supply and demand of ERCs that occur because the control technologies do not meet the target exactly; this varies from year to year and would tend to decrease the market price. On the other hand, they do not include any administrative, legal, or other transaction costs, which

Table 11.13 - Transfer Payments with ERC Sales and Leases

Control program	Transfer payments (10^6 dollars/year)[a]	Percent of control costs
1.5 lb. excess emissions	16 - 45	7 - 18
50% proportional reduction	17 - 42	8 - 20
1.8 lb. systems cap	18 - 44	11 - 27

[a]1982 dollars

would tend to increase it. The table shows that the range of transfer payments is similar under each of the scenarios, but these would represent an increasing proportion of the control costs as the scenarios become less stringent. A large portion of these payments would be borne by utility D, and the previous sections have shown that most would be in the form of leases or contingent sales.

12

Implementation Concerns

The concept of leasing ERCs is relatively new, and there obviously has not been much market experience that would identify implementation problems. Only one leasing arrangement has been reported in the literature,[1] and a contact with the agency involved indicates that a second similar case has also been approved.[2] Both of these involved the leasing of VOC ERCs from a shutdown plant owned by International Harvester in the Louisville, Kentucky, area. In the first case, General Electric leased 445 tons of VOC in 1982 and 280 tons of VOC in 1983 for a total cost of $60,000. General Electric was completing a $2 million modernization of a dishwasher manufacturing plant and needed ERCs until the project was finished. Pending approval of the regulatory agency, GE paid $5,000 for an option to lease the ERCs for the two years as noted above. A third year (1984) was also specified in the option, with GE retaining the right to lease up to 280 tons of VOC for $100 per ton; the third year option was not exercised, however. The option agreement was relatively simple—essentially a one page letter signed by both parties. A representative of GE suggested that the legal agreements between GE and International Harvester were simple because the ERCs were readily available, and because GE needed them for only a short period of time.[3]

In the second transaction, International Harvester leased 50 tons of VOC to Rohm & Haas for a plant expansion early in 1983. This was also a multiyear agreement, extending through 1985.[4] Rohm & Haas intended to use internal ERCs ultimately to provide the permanent offsets necessary for the project. It decided to lease ERCs from International Harvester to minimize any regulatory difficulties if the necessary plant modifications were not completed by the new unit's startup date. It also had an option to purchase the International Harvester ERCs in 1985 at the completion of the lease, should they be deemed necessary.

RESPONSE OF EMISSION BANKS

Since the two known leases of ERCs had taken place in an area with an active emissions bank, other locations with formal banks were contacted to determine if leasing had occurred, and also to ascertain the viewpoint of the regulatory agency toward leasing.

Three emission banks analyzed by the Environmental Defense Fund—the Bay Area (San Francisco) Air Quality Control District, Allegheny County (Pittsburgh), and Puget Sound District (Seattle)—were contacted, but no leasing transactions have occurred at these banks.[5] One representative referred to leasing as an "interesting concept"; none has any prohibitions against leasing. Another noted that it is "not an issue that has been addressed."

One bank made a clear distinction between different types of leasing arrangements:

1. *New Source Leases*. The regulatory agency would not want a new unit to lease ERCs to meet offset requirements. The spokesman noted that the unit's very existence would represent an argument in favor of continued operation, even if the lease expired. As he stated: "You would need a very strong argument to shut down a unit where capital investment has already been made."

2. *A Bridge for Compliance*. If a unit cannot meet a compliance deadline, then the firm could lease ERCs between the compliance deadline and actual installation (and initial operation) of the pollution control equipment. This is the type of lease employed by GE as discussed above. The spokesman termed this "a bit of a farce" since the emissions picture doesn't change at all. The pollutants associated with the banked ERCs, which were not being emitted, are still not being emitted, and the noncompliance firm is still out of compliance. Thus the end result was "hassling the out-of-compliance firm—charging them money—for no air quality benefit." He suggested instead the development of a compliance program to force firms to make a strong commitment, and then to make them push to meet the goals.

3. *Retiring Units*. The banking representatives agreed that a lease might be useful if a firm wanted to prolong the life of an existing facility. This bank recently had a plant close down rather than comply with the regulations, and a lease might have been applicable (although it was never analyzed, to his knowledge).

The viewpoints expressed by this emission bank representative raise a number of interesting implementation issues. First, the use of ERCs for new source offsets could become an important factor for utilities building new coal-fired units, particularly after the year 2000. If the agencies take the viewpoint that leasing is not acceptable for new source offsets, this enhances the likelihood of hoarding of ERCs by the individual utilities; that is, no utilities will sell at all, since they will hold all ERCs for their own future growth. This might actually enhance the use of leasing in the short term (to meet emission reduction requirements). For example, utilities C and E in the 1.5 lb. excess emissions control program will not make any ERCs available for sale, resulting in 100 percent leasing transactions. In the long run, however, after the initial reductions have been achieved, any new source offsets would probably have to rely strictly on internal transactions.

Second, the spokesman overstates the case somewhat against bridging for compliance. A banked ERC is not necessarily equal to a leased ERC, since the first represents a potential to emit, but the potential has already been spent in the second. Nonetheless, the argument is instructive because it points out once again the differences between ERCs created through overcontrol and those created through unit retirements.

There are two philosophies about the use of ERCs. The first suggests that air quality can be defined in terms of the baseline condition in the SIP. Any subsequent improvements—through overcontrol, plant retirements, unit curtailments, etc.—are all legitimate reductions as long as they are "surplus" with respect to the SIP baseline, or the plan to achieve the NAAQS that relies on this baseline. This approach has been used by the federal EPA.

A second approach suggests that there is a change in air quality over time associated with capital turnover. As newer, more efficient technologies are brought on-line and older, polluting equipment is retired, there will be a net benefit in air quality. ERCs from retiring units are therefore not necessarily equivalent to those from overcontrol, since they eventually would have occurred anyway. Several states have adopted this philosophy, and thus place restrictions on the use of ERCs associated with plant retirements. For example, these retirement ERCs may be applicable as offsets for a new unit moving in or a plant expansion, but they are not available in lieu of additional controls on existing units (in a bubble).

In the SEROP modeling case discussed above, the ERCs were obtained through added control technologies, and thus represent emission reductions that would not otherwise have occurred without the acid deposition control program. There were two exceptions, however—the generation shift for utility C and the turbine overhaul for utility J. The potential reluctance of agencies to grant credit for these has already been discussed.

But the effect of scheduled retirements is also a factor, because they become a part of the systems bubble. Six units are scheduled to retire over the 18 year period. These are used internally to meet required emissions limitations, but are not specifically offered as ERCs. They obviously affect the utilities' position with respect to compliance, however, and thus influence the number of ERCs bought and sold (or leased).

It is therefore clear that the manner in which the acid deposition control program is implemented can affect the generation of ERCs. If the states focus on permits for each plant within the bubble (rather than a total system permit), they might seek to make a distinction between scheduled retirements and units that are prematurely retired. The former's ERCs would be appropriate only for offsets of future units, while the latter's would presumably be useful as controls to achieve emissions reductions. This distinction was not made in the SEROP analysis, when all ERCs were treated as being equivalent. No premature unit retirements were addressed (as noted earlier, the utilities thought that this would occur only if the regulations were extremely onerous).

The issue has significance in the modeling analysis, since the General Electric–International Harvester case represents conditions similar to the replacement of capacity by utility G (with the exception that the time period for leasing would be much longer for the utility). In both cases, older, higher-emitting facilities would be replaced by newer, less-polluting units. The lease serves to "cover" the differences between the compliance date and installation of the new unit and appears to be one of the demand conditions most conducive to a leasing program. The emissions banking representative has reservations about using ERCs from retired units for that "cover," but would probably not object if these ERCs were obtained from overcontrol.

The final issue was a more favorable judgment concerning the potential for leasing to extend the life of an existing unit. This was the case for utility D, and it could potentially be used to prolong the life of the units scheduled for retirement. Given the high costs of replacement capacity and the recent attention to extending boiler life, this could become a relatively common occurrence in future years. The arguments above about the nature of available ERCs could still be raised, but it appears that control agencies would be receptive. Since the conditions are so conducive to leasing, it will probably be a major source of leasing demand. The use of large quantities of leased ERCs for extended periods of time (as in the case for utility D) might be opposed, however, by those who would prefer to see localized control measures despite their higher costs.

PROHIBITIONS

Despite the comments above concerning willingness to accept leasing, at least one jurisdiction has already prohibited it. Rule 210.3 of the

Kern County (CA) Air Pollution Control District (APCD) concerning Emission Reduction Banking states that "Until provisions are incorporated in this rule, leasing of banking certificates shall not be allowed." This rule was adopted in April 1983. A discussion with a Kern County representative indicated that the ERC leasing prohibition was not included because of a specific leasing problem or situation; indeed, the APCD was not aware of anyone who had leased ERCs, nor were they aware of anyone else who had prohibited leasing.[6] ("Most places haven't even thought of it yet.") He suggested that a typical problem might arise if the lessee puts in a new unit and was polluting, but the lessor does not meet his emission control requirements or sells the business.

When it was pointed out that this was no different than the type of enforcement problem in any external transaction, the representative agreed, stating that leasing was "just one more complicating step which we don't want." He went on to suggest that "banking certificates don't help the district—they only help industry." The Air Pollution Control District felt that there were a number of problems in their rules for New Source Review, and they didn't want the same types of problems with trading and banking, so they "drew up the regulations pretty tight." He thought that leasing arrangements could become quite complex legally.

While the rule has the qualifying statement "until provisions are incorporated in this rule," there are currently no plans for making such provisions in the foreseeable future.

It is not likely that outright prohibitions of this type will be a major implementation obstacle. More important, however, is that it indicates the reluctance that many pollution control agencies display toward emissions trading. To many persons, emissions trading is still a novelty. The leasing of ERCs is one more issue they would prefer not to address.

TAX AND FINANCIAL CONSIDERATIONS

Consider the scrubbers for units EA73-2 and EA75-3 listed in Table 11.2. Both of these scrubbers are cost-efficient in meeting the emissions requirement in the state, but the total reduction is not necessary for the individual utility. Hence ERCs are available from the scrubber on unit EA75-3. If the utility is going to finance these scrubbers, the assets associated with the scrubber on EA73-2 consist of the equipment itself. For the scrubber on EA75-3, however, there is not only the equipment but also the income stream associated with the leased ERCs. The tax and financial issues associated with these differences could become an important implementation issue, and could lead to some fundamental shifts from existing approaches.

The major tax and financing issues concerning ERCs have been discussed by Baker and Winslow.[7] Obviously, the treatment of taxes and financing varies not only from state to state but also from utility to utility (that is, investor-owned, cooperatives, or municipal units), and specific application of tax law must be addressed on a case-by-case basis. This is particularly true given the recent radical changes in tax law. Nonetheless, several general issues appear to be relevant concerning leased ERCs.

First, Baker and Winslow point out that the primary tax benefit to a firm leasing control equipment is the deductibility of rental payments. As noted in the beginning of Chapter 10, Baker and Winslow feel that many firms may decide that long term leasing arrangements are preferable to purchasing ERCs since the tax benefit of amortization deductions may not be available. Two concerns are raised about the amortization deductions: (a) the deductions for the cost of an intangible asset are allowable only to the extent that the ERC cost exceeds its salvage value, and (b) the firm must establish a useful life for the ERC. While the use of contingency sales might help establish the useful life criteria, the resale value will remain high. The salvage value cannot be amortized, but, as Baker and Winslow note, "can only be recouped for tax purposes when the ERC is sold. At such time it is reflected in determining the amount of taxable gain or loss realized on the sale."[8] The use of ERC leasing to avoid the amortization problem could therefore make leasing preferable to contingency sales.

On a more historical note, Baker and Winslow indicated that "industrial development bonds will not be available if the primary purpose of the bond issue is to finance the acquisition of an ERC."[9] Thus, utility D would not have been able to finance its required ERC through industrial development bonds, but presumably the annualized lease payments would not have been larger than the finance costs it would have had to make. The authors also addressed the impact of ERC creation on bond financing and noted that to the extent that the firm was able to recover all or part of its investment through the sale of an ERC, this portion of the investment might not have been included as part of a qualifying facility. Hence, creation of an ERC might have impeded financing for the project. They did not specifically address the question of lease rentals, but the "economic benefit" criteria used by the IRS could similarly have posed an impediment for ERC leasing arrangements under bond financing.

The "economic benefit" question was very tenuous, relying on a 1975 proposed Treasury regulation that was never finalized (but which nevertheless represented IRS policy). Baker and Winslow's discussion noted that problems with an ERC would have been caused only by an adverse IRS "economic benefit" determination, and there was some preliminary indication through letter rulings that this would not occur. Legislation was also introduced in Congress that would have modified the economic benefit formula to reduce the qualifying portion only when the facility generates a "net profit."[10]

A recent analysis by Raufer et al. addressed these concerns but has been rendered essentially moot by the tax law revisions enacted in 1986.[11] The analysis determined that leasing was preferable to the buying and selling of ERCs under the most probable 1985 tax conditions, which would allow the seller capital gains treatment on the income from the sale of the ERC, and would allow the purchaser an amortized deduction over the life of the ERC. The magnitude of the savings in a potential transaction between utilities C and D (with utility C using FGD to create the ERC) was $14 to $19 million from a total revenue requirement of $276 to $308 million. Note that this is the savings from leasing an ERC rather than buying/selling it, *not* the savings associated with ERCs over a command/control program.

Since the life of an ERC is more than one year, it is unlikely that an immediate deduction by the buyer would be allowed by the Internal Revenue Service; in the event that it were, leasing was not as attractive as a sale (by approximately $11 million).

The results were not changed by the level of tax-exempt industrial development bonds (IDBs) used to finance the scrubber. In the event that the IRS were to make a negative ruling on the question of the "economic benefit" of an ERC, and thus restrict IDB financing to that portion of the scrubber not associated with ERCs, leasing remained the preferred alternative for the most likely tax treatment.

Favorable results were also obtained in the situation where higher operating expenses were incurred (through the purchase of low-sulfur coal) to create the ERC. In this transaction, leasing required lower revenues than an ERC sale regardless of the tax positions taken by the ERC buyer or seller. One of the factors that played a role in this case, however, was the AAA rating of utility B employing the low-sulfur coal.

The tax changes of 1986 repealed both the IDBs for these types of control systems and Investment Tax Credits (ITCs), and also instituted less generous depreciation schedules. They also reduced the corporate tax level. These changes will have significant impacts on capital intensive systems such as scrubbers. The modifications seem likely to increase revenue requirements, and the savings associated with ERC leasing could therefore be even greater; once again, however, the tax implications must be addressed on a case-by-case basis.

Finally, given the current problems many utilities are having in the bond market because of their nuclear plant exposure and the large amounts of capital required for acid rain control, it is interesting to speculate on the applicability of more innovative financing techniques if financing were not specifically addressed in the acid deposition control legislation.

The success of third-party financing arrangements in the energy field and the dramatic shift in the 1970s toward the use of leveraged leasing could lead financing for pollution control equipment to move in a similar direction. A leveraged lease, for example, derives its name from the position of the lessor in the transaction: the lessor leverages its capital investment, borrowing the remainder from a third party and using the lessee's rental payments, the equipment, and the lessee's financial capacity as collateral. Leveraged leases are typically used for "big ticket" capital intensive items, and offer the lessee certain advantages, the primary one being lower costs.[12] Thus leveraged arrangements could be appropriate for pollution control equipment.

We have already noted some of the potential problems an ERC could have introduced for bond financing. The ERC appears to have just the opposite effect under a leveraged lease arrangement, however. By increasing the assets associated with the same loan costs, the lessor would have the opportunity to leverage even further or to provide greater collateral. This is not to say that the creation of an ERC alone would shift financing for pollution control equipment to a leveraged leasing arrangement. Fritch, for example, points out numerous problems involved in structuring a leveraged lease for a public utility, including problems associated with individual state public utility statutes, the Public Utility Holding Company Act of 1935 (is the lessee issuing a "security" requiring SEC approval?), and the Federal Power Act.[13]

Nor should leasing be viewed as a panacea. In former economic studies of leasing arrangements, utility executives have raised concerns about the inability to obtain a rate of return on leased property since it is not included in the rate base, and the adverse perception of rating agencies if leasing constitutes a large percentage of the firm's capitalization (one utility kept its percentage below a targeted 5%).[14] Nonetheless, ERC leasing may ultimately have an important impact on the financing of pollution control equipment in an acid deposition control program.

13 Conclusions

The difficulty in implementing a market based approach to pollution control, even under the reasonably favorable conditions and sophisticated infrastructure that exists in the United States, is apparent. The existing emissions trading market has been slow in developing, due to its constrained nature and also to other impediments such as uncertainty about future regulations. Electric utilities, which will shoulder a significant portion of any acid rain control emission reductions, have been affected by these factors. They have also been affected by utility-specific impediments such as rate-making concerns and the desire to save ERCs for customers, although these have not been quite as important. These general and utility-specific impediments have caused utilities to hoard ERCs and have contributed to making the ERC market very thin.

Emissions trading in an acid deposition program could be minimized by the proper emissions reduction allocation, but it is still likely to display many of the characteristics of the existing ETP program. The vast majority of transactions will continue to be internal. Even if allowed, there will probably be no significant usage of interfirm or interstate trades by electric utilities. Market constraints will continue to be important, although the easing of dispersion modeling constraints to allow systems bubbles will probably expand the use of internal emissions trading considerably. Utilities are strongly in favor of this trading option, which creates flexibility with little risk.

External trades will continue to face obstacles from the market impediments, as well as from other factors such as the utilities' organizational decision-making process that stresses strategic aspects of the control technology decision. External trades also depend on the generation of potential ERCs, through overcontrol or unit retirement. While the latter are viewed by utilities as potential ERCs (though not necessarily for sale), the utilities appear to be reluctant to release their overcontrol "cushion," and the regulatory agencies to grant credit for the overcontrol.

With these constraints in mind, we proposed ERC leasing as a mechanism that could enhance the viability of the market, allowing the utility to maintain future control of ERCs while using them in the market. We analyzed ERC leasing under three acid rain control scenarios in the state of Illinois to see whether a potential for using the mechanism exists. We were surprised at the strength of the results; ERC leasing was a *necessary component* in the market, because of the utilities' need for the ERCs in future years.

Utilities have hoarded ERCs, and have been unwilling to part with them regardless of price. The market analysis we conducted based ERC prices on the marginal costs of control technology. These prices do not take into account the differences between supply and demand of ERCs that occur because the control technologies do not meet the allowable target exactly. This varies from year to year and would tend to decrease the market price. On the other hand, administrative, legal, and other transaction costs are not included in the estimates and would tend to increase it.

More important, however, are the comments of one utility spokesman, who reviewed our ERC prices based on control costs, and wrote: "That's only half the issue—equally important is the value, which [you have] totally ignored." At least part of his concern about "value" refers to the strategic aspects of the ERC. Concern for system reliability and the uncertain requirement for future emission reductions are considerably more important to utility decision-makers than near-term cost-efficiency gains, and utilities would therefore tend to hoard the ERCs unless the price reflected these concerns as well as the control costs. This is, of course, one of the principal strengths of the leasing transaction; the utility generating the ERC would not have to include these "value" costs in pricing because it would still own the ERC.

A pure economist might argue about the "inventory clerk" nature of the supply-demand analysis, suggesting that a market clearing price will occur under *any* conditions. In the real world, however, a utility analyst who thinks about selling an ERC today, but holds back because he might not be able to find one in five years at any price, is much more likely to enter the market through leasing instead of selling (even at some hypothetically high price).

While leases are not a panacea, we believe that the leasing mechanism can enhance the market approach to pollution control for acid rain mitigation in the United States, and achieve emissions reduction at lower costs for ratepayers. Charles Schultze has noted that "the literature on the efficiency of markets and on the advantages of using market-like processes for social intervention almost never recognizes the problem of creating markets *de novo*."[1] Given the novelty of property rights associated with these types of market approaches, we believe that leasing could play a major role in implementing many other market-based systems as well.

Notes

CHAPTER 1

1. R.M. Alison, "The Earliest Traces of a Conservation Conscience."

2. T.M. Schmitz, "Pollution, Law, Science, and Damage Awards," p. 457, but questions about its authenticity are raised in P. Brimblecombe, "Attitudes and Responses Towards Air Pollution in Medieval England," p. 944.

3. Achieving compliance once offenders have been identified is a difficult and important task under any regulatory scheme. For a discussion of enforcement under the Clean Air Act, see N. de Nevers, "Enforcing the Clean Air Act of 1970," pp.14–21; and National Commission on Air Quality, *To Breathe Clean Air*, part 3, chapter 8.

4. A.C. Pigou, T*he Economics of Welfare*. See also A.V. Kneese and C.L. Schultze, *Pollution, Prices and Public Policy*; P. Burrows, *The Economic Theory of Pollution Control*; and J.J. Seneca and M.K. Tausig, *Environmental Economics*.

5. T.H. Tietenberg, "Transferable Discharge Permits and the Control of Stationary Source Air Pollution: A Survey and Synthesis," p. 391.

6. The PPBS approach was adopted under President Johnson in August 1965. For a critique of this system, see A. Wildavsky, *The Politics of the Budgetary Process*, chapter 6.

7. G. Majone, "Choice Among Policy Instruments for Pollution Control," p. 589.

8. Ibid., p. 610

9. G.F. Sandbach, *Environment, Ideology and Policy,* p. 78.

10. K. Coates, Ed., *Socialism and the Environment*; H. Stretton, *Capitalism, Socialism and the Environment*.

11. A. Gorz, *Ecology as Politics*, p. 18.

12. B. Devall and G. Sessions, *Deep Ecology*.

13. M. Douglas, "Environment at Risk," pp. 1273–1275; idem, *Purity and Danger*; A. Wildavsky, *Speaking Truth to Power*, chapter 8, pp. 184–203; and M. Douglas and A. Wildavsky, *Risk and Culture*.

14. The General Accounting Office (GAO), for example, notes that "in analyzing problems troubling controlled trading—and with its transition to a full scale market in mind, we concluded that EPA should promote controlled trading as an alternative, rather than as a complement (supple-

ment) to command and control." See General Accounting Office, *A Market Approach to Air Pollution Control Could Reduce Compliance Costs Without Jeopardizing Clean Air Goals*, p. 150.

15. E.B. Easton and F.J. O'Donnell, "The Clean Air Act Amendments of 1977: Refining the National Air Pollution Control Strategy," p. 947.

16. See Wildavsky's review in *Speaking Truth to Power*, chapter 8, of B. Ackerman, S. Rose-Ackerman, J.W. Sawyer, and D.E. Henderson, *The Uncertain Search for Environmental Quality*.

17. J. Lovelock, *Gaia*, p. 27.

18. D. Roe, *Dynamos and Virgins*.

19. F.D. Krupp, "New Environmentalism Factors in Economic Needs"; Environmental Defense Fund, *To Burn or Not to Burn: The Economic Advantages of Recycling Over Garbage Incineration for New York City*.

20. U.S. Council on Environmental Quality, *Annual Report on Environmental Quality for 1984*, Table A-19, pp. 614 ff.

21. R.A. Liroff, *Reforming Air Pollution Regulation: The Toil and Trouble of EPA's Bubble*.

22. Ibid., p. 136.

23. Ibid., p. 16.

CHAPTER 2

1. From 1974 to 1986 the utility industry committed through EPRI over $350 million for acidic deposition research and development activities, and 1986 plans were to spend about $180 million more over the next three years. See Science and Policy Associates, "Acidic Deposition Catalog."

2. The background pH is about 5.0. See World Resources Institute (WRI), *World Resources 1986*, p. 169.

3. On a global basis, about 50 percent of atmospheric sulfur is from natural sources.

4. WRI, *World Resources 1986*, pp. 161–66.

5. Ibid., p. 167.

6. Kenneth Rahn and Douglas Lowenthal of the University of Rhode Island Graduate School of Oceanography have shown (in work completed for DOE in 1986) that the ratio of several trace elements in the surface waters of the northeastern United States are such that the sources of the precursor emissions are largely from middle western states. Specifically, they find that 80 percent of the acid deposition in the Northeast is attributable to smokestacks from middle western emission sources. Similarly, they showed that large amounts of the acid rainfall in Canada are from a similar pattern of emission sources; telephone conversation with D. Lowenthal, November 18, 1986. See K.A Rahn and R. Heston, "Elemental Tracers of Source Regions of Contaminants in Precipitation," 5th Status Report, unpublished, for DOE Grant No. DE-F022-84PC71253, August 1, 1986. This work has not been formally published but is available from the

authors. The findings are also discussed in "Analysis Said to Pinpoint Sources of Acid Rain," *New York Times*, August 30, 1986. These latest findings render obsolete their earlier work, which sampled only low-altitude airborne particulates. The airborne particulate matter was significantly traceable to local sources in at least half the samples. The explanation for the anomalous earlier findings is that the higher the altitude of the sampled air, the greater the distance it had traveled. The precipitation of moisture from the atmosphere is from relatively high altitudes compared with the original sampling, and is laden with acidic materials from sources several hundred miles upwind. The most recent findings suggest that about 80 percent of the sulfates in rain and snow falling on the Northeast are probably from the Midwest.

7. WRI, *World Resources 1986*, p. 167.

8. "Developing the Options for Emissions Control," *EPRI Journal*, November 1983, p. 41; the entire issue is devoted to the acid rain problem.

9. EPA, "Acid Deposition: Current Knowledge and Policy Options," Appendix II, pp. 8, 10.

10. pH 4.2 to 4.6.

11. Vinegar has a pH of about 3. Battery acid has a pH of about 1, and the rain in Wheeling had a pH of about 1.5; see WRI, *World Resources 1986*, p. 168.

12. That are below a pH of about 5.

13. In general, wet and dry sulfate deposition models assume reductions (that is, hydrogen ion concentration reductions in surface and groundwater) will be related in a linear way to emission reductions. There is reason to believe that the assumption of linearity may approximate reality over larger regional scales and long enough time periods, but the potential exists for nonlinear results during episodic events or in restricted areas. Nevertheless, because of mass balance considerations, a given reduction in emissions must translate into a corresponding decrease in total deposited sulfur, and the preponderance of evidence is that the linearity assumption reasonably captures the effects. An NAS study in 1982 awkwardly concluded that on a scale the size of eastern North America, there is no evidence for a strong non-linearity in the relationships between long-term average emissions and deposition.

14. EPA, *Position Paper on Regulation of Atmospheric Sulfates*.

15. E.G. Wetstone, "The Need for a New Regulatory Approach"; J. Ostrov, "Interboundary Stationary Source Pollution—Clean Air Act Section 126 and Beyond."

16. HR 3400, 98th Congress.

17. A.H. Gilbert, "The Economics of Acid Rain: An Invisible Hand of Control," pp. 85–90; E.F. Renshaw, "A Tax Subsidy Approach to Acid Rain Abatement," pp. 47–48.

18. ICF, Inc., *Alternative Strategies for Reducing Utility SO_2 and NO_x Emissions*.

19. See Kneese and Schultze, *Pollution, Prices, and Public Policy*, p. 99.

20. Kneese and Schultze, ibid., cite one emission fee study that "could achieve the same environmental objectives at only 10 percent of the costs of the uniform cutback method."

21. The National Commission on Air Quality, for example, examined the feasibility of using an emissions fee to achieve the ambient ozone standard, instead of implementing Reasonably Available Control Technology (RACT) on volatile organic compound (VOC) emissions in the Twin Cities, Minnesota area. See R.K. Raufer, L.G. Hill, and M.E. Samsa, "Emission Fees and TERA: An Evaluation of Policy Alternatives in the Twin Cities," pp. 839–45. For a fee based on an index, see A.M. Schneider, "An Effluent Fee Schedule for Air Pollutants Based on PINDEX."

22. T.D. Crocker and B.A. Forster, "Decision Problems in the Control of Acid Precipitation: Non-convexities and Irreversibilities," pp. 31–37.

23. Ibid., p. 34.

24. D.M. Violette, "Comment on Decision Problems in the Control of Acid Precipitation: Nonconvexities and Irreversibilities." Also see the Author's reply, same issue.

25. A. Rosencranz, "Economic Approaches to Air Pollution Control," p. 27.

26. Ibid.

27. See, for example, Air Management Services, *Feasibility Study: A Fee/Subsidy System for Controlling Sulfur Dioxide Emissions in Philadelphia*.

28. See C.S. Russell, "What Can We Get from Effluent Charges?" pp. 155–80; R.J. Tobin, "Is Regulation Outdated? Achieving Clean Air with Alternatives to Regulation"; L.M. Wenner, "Pollution Control," pp. 47–65; Ackerman et al., *The Uncertain Search*; Majone, "Choice Among Policy Instruments"; and Raufer et al., "Emission Fees."

29. The Mitchell bill would have allowed "trading of emission reduction requirements and actual reductions on a State or regional basis, for which States and Administrator (of EPA) are authorized to establish emission reduction banks or brokerage institutions to facilitate such trading." Emissions trading is relevant in an acid deposition control program not only because of the emissions reductions required; it has also been the mechanism proposed to handle growth. Some bills, for example, would require offsets for any emission increases from new plants as well.

30. Liroff, *Reforming Air Pollution Regulation: The Toil and Trouble of EPA's Bubble;* and T.H. Tietenberg, *Emissions Trading: An Exercise in Reforming Pollution Policy*. Liroff's earlier work, *Air Pollution Offsets: Trading, Selling and Banking*, is also relevant.

31. While Tietenberg devotes an entire chapter, for example, to the temporal dimension of emissions trading, he never mentions the principal concern of this book—the effect of future internal ERC needs on the supply in the market.

32. 41 FR 55524.

33. B. Yandle, "The Emerging Market in Air Pollution Rights," p. 26.

34. Council on Wage and Price Stability, "Council Endorses EPA Air Emission Trade-Off Policy."

35. National Commission on Air Quality, *To Breathe Clean Air.*

36. J.S. Hoffman, "Economic Advantages of Emissions Banking Systems."

37. 44 FR 3280.

38. ICF, Inc., *Emission Reduction Banking Manual.* One aspect of the emissions bank is that the bank does not increase the size of the holdings, as usually happens with currency. Instead, the holdings are often discounted.

39. 51 FR 43814.

40. 44 FR 71780.

41. M.H. Levin, "The Supreme Court's Bubble Decision: What It Means," pp. 10–11.

42. Netting is another controversial element of the ETP, which Liroff defines as a "procedure whereby dischargers can avoid some of the Clean Air Act's most demanding and technological requirements." It allows a credit for other nearby reductions; by ensuring that the "net" increase in emissions is insignificant, the source can avoid full new source review requirements. See Liroff, *Reforming Air Pollution*, p. xxi.

43. Telephone conversation with B. Elman, EPA's Regulatory Reform Staff, December 17, 1984.

44. EPA, *Fact Sheet on Emission Trading Policy.*

45. W. Vivian and W. Hall, *An Examination of U.S. Market Trading in Air Pollution Offsets*, chap. 2.

46. Ibid., chap. 4.

47. Ibid., chap. 2.

CHAPTER 3

1. There were exceptions, of course; D.T. Stathos and M.S. Treitman ("Using Private Market Incentives for Air Cleanup," pp. 13–21), R.W. Hahn and R.G. Noll ("Barriers to Implementing Tradable Air Pollution Permits: Problems of Regulatory Interactions," pp. 63–91), and J.M. Gerard ("The Effects of Public Utility Regulation on the Efficiency of a Market for Emission Permits,") have identified utility-specific issues, which we will return to in the text below.

2. Reasons for this are discussed in later sections.

3. Vivian and Hall, *An Examination*, chap. 2.

4. J.E. Norco and R.K. Raufer, "Experience with an Emissions Trading Program."

5. S.A.B. Blackman and W.J. Baumol note, "The offset policy has only produced a very rudimentary approximation to an organized market" ("Modified Fiscal Incentives in Environmental Policy", p. 426). The General Accounting Office echoes the sentiment: "EPA's controlled trading policies are a limited market approach" (*A Market Approach*, p. 150).

6. GAO, *A Market Approach*, Appendix II, pp. 117–24.

7. "NSPS Bubble Issues"; J.P. Bromberg, *Bubbling New Sources with Existing Sources: An Examination of the Environmental and Economic Benefits of an Integrated Approach to Emissions Control.*

8. EPA, *Fact Sheet.*

9. L.T. Kontnik, "The Evolving Emissions Trading Program: Fish or Fowl?," p. 5.

10. Telephone conversation with Pennsylvania utility spokesman who was promised confidentiality, March 28, 1983.

11. "Pennsylvania Air Chief Says EPA has 'Oversold' Air Bubble Policy," *Inside EPA*, December 11, 1981.

12. GAO, *A Market Approach*, p. 103.

13. D. Bass, "Job Creation and Clean Air: Can a City Achieve Both?," pp. 5–8; L.P. Malone, "Air Quality and Development: How They Can Complement Each Other," pp. 23–27.

14. GAO, *A Market Approach*, p. 98.

15. R. Males, "SURE Findings and Conclusions," p. 48.

16. B. Yandle, "*The Emerging Market*," p. 28.

17. C.S. Russell, "Controlled Trading of Pollution Permits," p. 27.

18. Council on Wage and Price Stability, "Council Endorses," p. 8.

19. These were issued in November 1986. The previous April 1982 policy statements did "not contain the detailed guidance that many states will need in developing banking systems," according to Kontnik ("Emissions Trading Program," p. 4).

20. GAO, *A Market Approach*, p. 123.

21. E.C. Breedlove, "Growth with Clean Air: An Analysis of the Emission Offset Policy," chap. 5, p. 2.

22. J.L. Landau, "Who Owns the Air? The Emission-Offset Concept and Its Implications," pp. 575–600; "Emission Offset Banking: Accommodating Industrial Growth with Air Quality Standards," pp. 937–64; and ICF, Inc., *Concept Paper: Emission Reduction Banking and Trading Systems.*

23. ICF, *Concept Paper.*

24. Kontnik, "Emissions Trading Program," p. 5.

25. The overall tax treatment of ERCs was examined in a report by Baker and Winslow, while Stathos and Treitman have addressed the tax implications from a utility's viewpoint. Both publications consider items such as tax-exempt bonds, basis, depreciation, investment tax credits, and capital gains or losses. Stathos and Treitman note that the Internal Revenue Service has yet to make definitive policy declarations in these areas, and they suggest that "the most effective way to reduce tax risks associated with (the ETP) is to request a private letter ruling" ("Private Market Incentives," p. 20). This could be quite time-consuming, however. See E.P. Baker and P.H. Winslow, *Tax Considerations Related to the Creation, Financing, Banking, Use and Disposition of ERCs in Controlled Trading Approaches to Air Pollution.*

26. Stathos and Treitman, "Private Market Incentives," p. 18.

27. GAO, *A Market Approach*, pp. 117–24.

28. E. Breedlove, "Growth with Clean Air." As one example, Breedlove cites one utility that "feared that the additional pollution control that they voluntarily placed on only some of their generators would someday be required on all of their generators" (chap. 3, p. 17). B.J. Calder echoes the concern that "the credit itself might tend to foster regulatory action" (*Development of an Offset Market in the California South Coast Area: A Marketing Research Study*, p. 4).

29. Vivian and Hall, *An Examination*, chap. 4.

30. Telephone conversation with R. Beck, Edison Electric Institute, January 3, 1983.

31. Peter Everett, legal counsel from a utility trade group on the acid deposition issue, suggested in a telephone conversation on January 3, 1983, that leasing may enable utilities to maintain control over their SO_2 ERCs in the future.

32. GAO, *A Market Approach*, p. 124.

33. J.D. Foster and M.C. Weiss, "Indirect Offsets—A Supplemental Program to Facilitate Interfirm Trades in Emission Reductions."

34. GAO, *A Market Approach*, p. 96.

35. Ibid., p. 124.

36. Vivian and Hall, *An Examination*, chap. 9. Also see P. Shabecoff, "U.S. Backs Trading in 'Credits' Allowing New Plants to Pollute"; and M.M. Hamilton, "EPA Promotes Private Method of Air Cleanup."

37. EPA, *Brokering Emission Reduction Credits: A Handbook*.

38. "Save Pollution Credits for a Rainy Day," *Chemical Engineering*, October 23, 1978, pp. 106–10.

39. Raufer et al., "Emission Fees," p. 844.

40. R.A.N. McLean, "The Relative Contributions of Sulfuric and Nitric Acids in Acid Rain to the Acidification of the Ecosystem: Implications for Control Strategies," pp. 1184–87.

41. E.M. Trisko, "Acid Precipitation: Causes, Consequences, Controls," p. 23.

42. Telephone conversation with J. McKenzie, Pacific Gas & Electric Co., January 17, 1983.

43. GAO, *A Market Approach*, p. 96.

44. Stathos and Treitman, "Private Market Incentives," pp. 20–21.

45. R.W. Hahn and R.G. Noll, "Barriers to Implementing," pp. 84–85.

46. Ibid., p. 88.

47. Gerard, "Public Utility Regulation," p. 196.

48. Ibid.

49. R.E. Balzhiser, "Energy Options to the Year 2000," p. 90.

50. S.A. Watson, "Banking and Emission Offset Policy: A Market Approach to Pollution Control," p. 59.

51. Environmental Law Institute, "Comparison of Operating Emission Reduction Banks."

52. "Public Meeting (24 January 79) to Consider a Proposed Model New Source Review Rule," California Air Resources Board, issued 3 January 79. Cited in Vivian and Hall, "*An Examination*," chap. 3.

53. Telephone conversation with R. Beck, Edison Electric Institute, January 3, 1983.

54. R.W. Crandall, "The Use of Environmental Policy to Reduce Economic Growth in the Sun Belt: The Role of Electric Utility Rates," p. 125.

55. Ibid., p. 126.

56. Stathos and Treitman, "Private Market Incentives," p. 18.

57. "PG&E Plans Unique Power Mix for 1980s."

58. K.M. O'Brien and B. Euser, "Solar Energy and the Search for Emission Offsets," pp. 1077–93.

59. Vivian and Hall, *An Examination*, chap. 7.

60. Ibid., chap. 3.

61. B. Garelick, "The Effect of Economic and Environmental Factors on Cogeneration Development in California."

62. R.C. Rittenhouse, "Environmentally Speaking," p. 40.

CHAPTER 4

1. J.V. Barker, "Electric Energy Brokering: An Explanation and Status Report," pp. 28–36.

2. Unpublished ICF data obtained from J. Austin, U.S. EPA, February 1983.

3. ICF, Inc., *Preliminary Analysis of the Mitchell Bill*, p. 3.

4. General Accounting Office, "Analysis of the Acid Rain Proposal Approved by the U.S. Senate Committee on Environment and Public Works on July 22, 1982," p. 8.

5. ICF, Inc., "Analysis of a Senate Emission Reduction Bill S-3041" (draft).

6. Telephone conversation with S.E. Elstein, General Accounting Office, June 14, 1983.

7. GAO, *Analysis of the Acid Rain Proposal*, p. 9.

8. Ibid.

9. GAO, *A Market Approach*.

10. Vivian and Hall, *An Examination*, chap. 3.

11. Ibid., chap. 7 (case SCO2).

12. Ibid., chap. 7 (case PA11).

13. Telephone conversation with A. Courtney, Commonwealth Edison Co., January 14, 1983.

14. Environmental Law Institute, *Institutional Aspects of Transported Pollutants: An Examination of Strategies for Addressing Long Range Air Pollution Problems*.

15. Ibid.

16. Massachusetts State Acid Rain Program, *Emissions Trading Within Acid Deposition Control: An Evaluation of Program Implementation and Regulatory Issues*.

17. A. Pasztor, "Market Booms for 'Rights' to Pollute."

18. See ICF, Inc., *Alternative Strategies for Reducing Utility SO_2 and*

NO_x Emissions; Teknekron Research, Inc., *Electric Utility Emissions: Control Strategies and Costs*; Environmental Law Institute, *Institutional Aspects*; and Office of Technology Assessment, *The Regional Implications of Transported Air Pollutants: An Assessment of Acidic Deposition and Rain*. The DOE/EPA study conducted by ICF, Inc., examined 22 emission reduction options; Teknekron Research, Inc., examined 12 strategies; the Environmental Law Institute (ELI) studied the institutional impacts of 9 options; OTA examined 11. Additional options have been studied by private sector organizations, such as the Edison Electric Institute and the National Clean Air Council/National Wildlife Federation.

19. More recent efforts have focused on an improved version of USM known as AUSM (the *A*dvanced *U*tility *S*imulation *M*odel). See C.J. Badger and H. Ojalvo, *The State-Level Advanced Utility Simulation Model: Documentation of System Design*.

20. B.G. Silverman, "Acid Rain and a Comparative Analysis of Two Large Scale Electric Utility Models," pp. 1031–42.

21. Ibid.

22. Hahn and Noll, "Barriers to Implementing," p. 91.

23. Ibid., p. 72.

24. B.A. Ackerman and W.T. Hassler, *Clean Coal/Dirty Air, or How the Clean Air Act Became a Multibillion Dollar Bail-out for High Sulfur Coal Producers and What Should Be Done About It*; and P. Navarro, "The 1977 Clean Air Act Amendments: Energy, Environmental Control, and Distributional Impacts," pp. 121–46.

25. OTA, *The Regional Implications*.

26. ICF, Inc., *Alternative Strategies*.

27. M.J. Roberts and J.S. Bluhm, *The Choices of Power*, pp. v–vi.

28. Ibid., p. vi.

29. Ibid., p. 17.

30. Ibid., p. 63.

31. Ibid., p. 119.

32. Certain utilities have already taken major steps in this direction. Wisconsin Power & Light, for example, has developed a computer program called SEROP (SO_2 Emission Reduction Optimization Program), which is designed to provide marginal cost data that is "objective, accurate, capable of withstanding considerable public scrutiny, and in general perceived as being unbiased and truthful." See G.E. Jablonka, V.N. Endres and D.G. Neumeyer, "SO_2 Emission Reduction Optimization Program (SEROP)"; "Utility's New SEROP Model Provides Optimized Emission Reduction Strategies." The firm has used the program to keep Wisconsin lawmakers informed about the impact of legislation, noting for example that the Mitchell Bill would have have required a 12 percent increase in revenue, reducing emissions 88,000 tons/year for the firm at an average cost of \$638/ton of SO_2. While this in no sense implies that Wisconsin Power & Light is in favor of an acid deposition control program, it is apparent that the firm has already given considerable thought to the external pressures it is facing, is aware of the impacts that the external

pressure will have on company operations, and with detailed, systemwide marginal cost data is in a good position to assess the potential savings associated with emissions trading.

33. C.F. Jaedicke, "Strategic Planning: A View from the Bottom Up," p. 52.

34. See, for example, N.H. Highton and M.G. Webb, "Sulphur Dioxide from Electricity Generation: Policy Options for Pollution Control," pp. 61–76.

35. Ibid., p. 74.

36. Roberts and Bluhm, *The Choices of Power*, p. 357.

37. Ibid., p. 35.

38. Ibid., p. 338.

39. Ibid., pp. 336–37.

CHAPTER 5

1. That is, West Virginia at 82 percent and Wisconsin at 72 percent.

CHAPTER 6

1. Vivian and Hall, *An Examination*.

2. See 45 FR 5616.

3. Central Illinois Public Service, Inc., *Clean Air Act, New Source Performance Standards, Recommendation of the New Source "Bubble" Concept (An Alternate Emission Reduction Option)*.

4. Pennsylvania Electric Co., "Continuous Emission Offsets."

5. 40 CFR 52.1881.

6. S.N. Putta, "Extension of Bubble Policy for Expanding the Use of Coal in New York Utilities." Putta's paper notes that the opinions are his, and do not represent the policies of the New York State Department of Public Service.

7. Interview with representative of Wisconsin Public Service Commission, April 7, 1983.

CHAPTER 8

1. Wetstone, "A New Regulatory Approach"; Ostrov, "Interboundary Stationary Source Pollution."

2. Kontnik, "Emissions Trading Program," p. 6.

CHAPTER 10

1. Baker and Winslow, *Tax Considerations*.

2. Ibid., p. 47.

3. Gerard, "Public Utility Regulation."

4. Jablonka et al., "SO_2 Emission Reduction."

5. D.G. Neumeyer, "Analyzing Emission Controls for Power Plants."

6. National Economic Research Associates, Inc. (NERA), "A Report on the Results from the Edison Electric Institute Study of the Impacts of the Senate Committee on Environment and Public Works Bill in Acid Rain Legislation (S. 768)."

7. Ibid., Appendix A; and Jablonka et al., "SO_2 Emission Reduction."

8. U.S. Dept. of Energy, *Inventory of Power Plants in the United States: 1982 Annual.*

9. Illinois Department of Energy and Natural Resources, *Illinois Energy Plan, Vol. 1: An Overview of Trends in the Illinois Electric Utility Industry.*

10. R.K. Raufer and S.L. Feldman, *Analysis of the Potential for ERC Leasing Transactions Under Various Acid Deposition Control Requirements.*

11. NERA, "Edison Electric Institute Study."

12. ICF, Inc., "Senate Emission Reduction Bill."

13. EPA: "Acid Deposition: Current Knowledge and Policy Options."

14. J.C. Molburg and E.S. Rubin, "Air Pollution Control Costs for Coal-to-Electricity Systems."

15. W.M. Vatavuk and R.B. Neveril, "Estimating Costs of Air Pollution Control Systems, Part XI: Estimate the Size and Cost of Baghouses"; J. Reynolds, S. Kreidenweis and L. Theodore, "Results of a Baghouse Operation and Maintenance Survey in Industry and Utility Coal-Fired Boilers."

16. R.D. Doctor, J.L. Anderson, D.B. Garvey, C.D. Livengood, and P.S. Farber, *The Impact of Coal Cleaning as a Sulfur Reduction Strategy in the Midwest.*

17. Rogers, Golden, and Halpern, Inc., *New Jersey Energy Facility Development Potential Study.*

18. R.C. Rittenhouse, "Upgrading and Maintaining Coal Handling Equipment."

CHAPTER 11

1. D.G. Streets, D.A. Knudson, and J.D. Shannon, "Selected Strategies to Reduce Acidic Deposition in the U.S."

2. ICF, Inc., "Senate Emission Reduction Bill."

CHAPTER 12

1. "The Buying and Selling of Pollution," p. 25.

2. Telephone conversation with M. DeBusschere, Jefferson County Air Pollution Control, Louisville, Kentucky, Sept. 29, 1983.

3. Telephone conversation with L. Vick, General Electric Co., Louisville, KY, Jan. 24, 1984.

4. Telephone conversation with D. Eagle, Rohm and Haas, Louisville, KY, Jan. 25, 1984.

5. Telephone conversations with J. Anderson, Puget Sound Air Pollu-

tion Control Agency, Nov. 22, 1983; S. Hill, Bay Area AQCD, Nov. 22, 1983; and C. Getz, Allegheny County Health Dept., Nov. 23, 1983.

6. Telephone conversation with T. Paxson, Kern County APCD, Nov. 8, 1983.

7. Baker and Winslow, *Tax Considerations*.

8. Ibid., p. 47.

9. Ibid., p. 56.

10. R.P. Feyer, "Governmental Incentives for Acquisition of Air Pollution Control Equipment."

11. R.K. Raufer, E.C. Bodmer, and J.A. Jaksch, "Tax and Financial Implications of ERC Leasing by Electric Utilities in Illinois."

12. R.A. Zanoni and A.J. Robison, "Leasing—A Preferred Alternative for Utilities?", pp. 35–39.

13. B.E. Fritch, "Leveraged Leasing."

14. The Conference Board, *Leasing: Experience and Expectations*.

CHAPTER 13

1. C.L. Schultze, *The Public Use of Private Interest*.

Bibliography

Ackerman, B.A.; Rose-Ackerman, S.; Sawyer, J.W.; and Henderson, D.E. *The Uncertain Search for Environmental Quality*. New York: The Free Press, 1974.

Ackerman, B.A., and Hassler, W.T. *Clean Coal/Dirty Air, or How the Clean Air Act Became a Multibillion-Dollar Bail-out for High Sulfur Coal Producers and What Should Be Done About It*. New Haven: Yale University Press, 1981.

Air Management Services, Department of Public Health, Philadelphia, Pa.; *Feasibility Study: A Fee/Subsidy System for Controlling Sulfur Dioxide Emissions in Philadelphia: Vol.1; Executive Summary*. December 1981.

Alison, R.M. "The Earliest Traces of a Conservation Conscience." *Natural History* 90:5 (May 1981).

Arey, D.G.; Grosskopf, S.; and Provenzano, G. *The Prospect for Controlled Trading of Air Pollution Rights in the Metro-East Area of Illinois*. Illinois Institute of Natural Resources Doc. No. 81/26, August 1981.

Atkinson, S.E. "Marketable Pollution Permits and Acid Rain Externalities." *Canadian Journal of Economics* 26:4 (November 1983).

Badger, C.J., and Ojalvo, H. *The State-Level Advanced Utility Simulation Model: Documentation of System Design*. University of Illinois, Prepared for U.S. EPA, November 1983.

Baker, E.P., and Winslow, P.H. *Tax Considerations Related to the Creation, Financing, Banking, Use and Disposition of ERCs in Controlled Trading Approaches to Air Pollution*. EPA Report #230-01-82-001, January 1982.

Balzhiser, R.E. "Energy Options to the Year 2000." *Chemical Engineering*, January 3, 1977.

Barker, J.V. "Electric Energy Brokering: An Explanation and Status Report." *Public Utilities Fortnightly* 109:3 (February 4, 1982).

Bass, D. "Job Creation and Clean Air: Can a City Achieve Both?" *Environmental Report*. National League of Cities, February 19, 1979.

Blackman, S.A.B.; and Baumol, W.J. "Modified Fiscal Incentives in Environmental Policy." *Land Economics* 56:4 (November 1980).

"Blowing Bubbles at EPA." *Regulation*, March/April 1979.

Brady, G.L., and Cunningham, R.D. "The Economics of Pollution Control in the U.S." *AMBIO* 10:4 (1981).

Breedlove, E.C. "Growth with Clean Air: An Analysis of the Emission Offset Policy." August 1979 (unpublished).

Brimblecombe, P. "Attitudes and Responses Towards Air Pollution in Medieval England." *Journal of the Air Pollution Control Association* 26:10 (October 1976).

Bromberg, J.P. *Bubbling New Sources with Existing Sources: An Examination of the Environmental and Economic Benefits of the Integrated Approach to Emission Control.* Prepared for Pennsylvania Electric Co. and New York State Electric and Gas Corp., December 9, 1980.

Budd, W.W. "A Systems Approach to Acid Deposition: Science, Institutions and Policy." Ph.D. dissertation, University of Pennsylvania, 1984.

Burrows, P. *The Economic Theory of Pollution Control.* Cambridge: MIT Press, 1980.

"The Buying and Selling of Pollution." *Chemical Engineering* 89:5 (March 8, 1982).

Calder, B.J. *Development of an Offset Market in the California South Coast Area: A Marketing Research Study.* Prepared for EPA Regulatory Reform Group, Washington, D.C., March 1982.

Catalano, L., and Makansi, J. "Acid Rain: New SO_2 Controls Inevitable." *Power*, September 1983.

Central Illinois Public Service Co., Inc. *Clean Air Act, New Source Performance Standards: Recommendation of the New Source "Bubble" Concept (Alternate Emission Reduction Option).* Springfield, Ill., April 24, 1981.

Clark, T.B. "New Approaches to Regulatory Reform—Letting the Market Do the Job." *National Journal*, August 11, 1979.

Coates, K. *Socialism and the Environment.* Nottingham: Spokesman, 1979.

Conference Board (The). *Leasing: Experiences and Expectations.* Report No. 791, 1980.

Costle, D.M. "The Bubble: Letting Industry Do the Job." *Pollution Engineering*, February 1980.

Council on Wage and Price Stability. "Council Endorses EPA Air Emission Trade-Off Policy." Washington, D.C., March 7, 1977.

Cowling, E.B. "Acid Precipitation in Historical Perspective." *Environmental Science and Technology* 16:2, February 1982.

Crandall, R.W. "The Use of Environmental Policy to Reduce Economic Growth in the Sun Belt: The Role of Electric Utility Rates." In M.A. Crew, *Regulatory Reform and Public Utilities.* Lexington, Mass.: Lexington Books, 1982.

Crocker, T.D., and Forster, B.A. "Decision Problems in the Control of Acid Precipitation: Non-convexities and Irreversibilities." *Journal of the Air Pollution Control Association* 31:1 (January 1981).

Dales, J.H. *Pollution, Property and Prices.* Toronto: University of Toronto Press, 1968.

DeFazio, J. "Offset Policy Memorandum: Cases to Date." Prepared for EPA Office of Enforcement, Washington, D.C., 1978.

del Calvo y Gonzalez, J.A. "Markets in Air: Problems and Prospects of Controlled Trading." *Harvard Environmental Law Review* 5:2 (1981).

de Nevers, N. "Enforcing the Clean Air Act of 1970." *Scientific American* 228:6 (June 1973).

Devall, B., and Sessions, G. *Deep Ecology*. Salt Lake City: Gibbs M. Smith, 1985.

Diemer, J.S. "Tradable Permits for the Control of Sulfur Dioxide Emissions in the Northeast Corridor: An Illinois Case Study." Master's thesis, University of Illinois, 1984.

Doctor, R.D.; Anderson, J.L.; Garvey, D.B.; Livengood, C.D.; and Farber, P.S. *The Impact of Coal Cleaning as a Sulfur Reduction Strategy for the Midwest*. Argonne National Labs., ANL/ECT-TM-7, EPA 905/2-83-001, June 1983.

Doniger, D.D. "The Dark Side of the Bubble." *The Environmental Forum* 4:3 (July 1985).

———. "The Bubble on the Cusp." *The Environmental Forum* 4 (March 1986).

Douglas, M. *Purity and Danger*. London: Routledge & Kegan Paul, 1966.

———. "Environment at Risk." (London) *Times Literary Supplement*, October 30, 1970.

Douglas, M., and Wildavsky, A. *Risk and Culture*. Berkeley: University of California Press, 1982.

Downing, D. *Air Pollution Emissions Offsets in the South Coast Air Basin: Institutional Designs to Promote Trading*. (preliminary draft), SRI International, Menlo Park, California, February 1982.

Easton, E.B., and O'Donnell, F.J. "The Clean Air Act Amendments of 1977: Refining the National Air Pollution Control Strategy." *Journal of the Air Pollution Control Association* 27:10 (October 1977).

"Emission-Offset Banking: Accommodating Industrial Growth with Air Quality Standards." *University of Pennsylvania Law Review* 128:4 (April 1980).

Environmental Defense Fund. *To Burn or Not To Burn: The Economic Advantages of Recycling Over Garbage Incineration for New York City*. New York, August 1985.

Environmental Law Institute. *Institutional Aspects of Transported Pollutants: An Examination of Strategies for Addressing Long Range Air Pollution Problems*. Prepared for National Commission on Air Quality, Washington, D.C., February 18, 1981.

———. "Comparison of Comprehensive Emission Trading Rules." Washington, D.C., April 1982 (unpublished).

———. "Comparison of Operating Emission Reduction Banks." Washington, D.C., April 1982 (unpublished).

Environmental Protection Agency (U.S.) (EPA). *Position Paper on Regulation of Atmospheric Sulfates*. Washington, D.C., 1975.

———. *Brokering Emission Reduction Credits: A Handbook*. Regulatory Reform Staff, Controlled Trading Project (PM-223), Washington, D.C., January 1981.

———. *Fact Sheet on Emission Trading Policy*. Washington, D.C., November 19, 1986.

———. "Acid Deposition: Current Knowledge and Policy Options." Briefing Document for the Administrator by the Acid Deposition Task Force, Washington, August 1, 1983.

"EPA's New Bubble: Not Everyone's Delighted." *Chemical Engineering* 89:15 (July 1982).

"Expanded Market Incentives Will Clean Air at Reduced Costs, Says GAO Report." *Journal of the Air Pollution Control Association* 32:6 (June 1982).

Feyer, R.P. "Governmental Incentives for Acquisition of Air Pollution Control Equipment." Presented at the APCA Specialty Conference on Economic Incentives for Clean Air, San Francisco, Calif., January 1981.

Foskett, W.H. *Emission Offset Policy at Work: A Summary Analysis of Eight Cases, Vol. 1*. Washington, D.C.: Performance Development Institute, April 1980.

Foskett, W.H.; Klaus, D.M; and Haberle, J. *Balancing the Objectives of Clean Air and Economic Growth: Regulated Markets in Emission Reductions*. Washington, D.C.: The Urban Institute, January 15, 1979 (issued June 1979).

Foster, J.D. "The Size and Price of Emission Offsets." Paper No. 78-12.5, presented at the 71st Annual Meeting of the Air Pollution Control Association, Houston, Texas, June 1978.

———. "Cleaning the Air for a Growing Economy." Paper No. 79-1.4, presented at the 72nd Annual Meeting of the Air Pollution Control Association, Cincinnati, Ohio, June 1979.

Foster, J.D., and Weiss, M.C. "Indirect Offsets—A Supplemental Program to Facilitate Interfirm Trades in Emission Reductions." Paper No. 81-52.2, presented at the 74th Annual Meeting of the Air Pollution Control Association, Philadelphia, Pennsylvania, June 1981.

Fritch, B.E. "Leveraged Leasing." Chapter 3 in B.E. Fritch and A.F. Reisman, eds., *Equipment Leasing-Leveraged Leasing*. New York: Practicing Law Institute, 1977.

Garelick, B. "The Effect of Economic and Environmental Factors on Cogeneration Development in California." Paper 83-29.7, presented at 76th Annual Meeting of the Air Pollution Control Association, Atlanta, Georgia, June 1983.

Garvey, D.B.; Moser, S.V.; Streets, D.G.; Speciner, T.A.; Dauzvardis, E.V.; Habegger, L.J.; Kellermeyer, D.A.; and Levine, E.P. "Nonattainment of National Ambient Air Quality Standards: Implications for Energy Policy—First Year Summary Report," Argonne National Laboratory, September 30, 1978 (unpublished).

General Accounting Office (U.S.). *A Market Approach to Air Pollution Control Could Reduce Compliance Costs Without Jeopardizing Clean Air Goals*. PAD-82-15. Washington, D.C., March 23, 1982.

———. *Analysis of the Acid Rain Proposal Approved by the U.S. Senate Committee on Environment and Public Works on July 22, 1982*. Washington, D.C., September 24, 1982.

Gerard, J.M. "The Effects of Public Utility Regulations on the Efficiency of a Market for Emission Permits." In G.R. Cass, R.W. Hahn, and R.G. Noll, *Implementing Tradable Emissions Permits for Sulfur Oxides Emissions in the South Coast Air Basin: Final Report to the California Air Resources Board*. Vol. II, June 30, 1982.

Gilbert, A.H. "The Economics of Acid Rain: An Invisible Hand of Control." *International Journal of Environmental Studies* 18:2 (1982).

Gorz, A. *Ecology as Politics*. Boston: South End Press, 1980.

Green, L. "Sulfur Emission and Acid Rain: A Least-Cost Approach to the Problem." *Public Utilities Fortnightly* 112:12 (December 8, 1983).

Greene, R. "Selling Dirt." *Forbes* 129:11 (May 24, 1982).

Hahn, R.W., and McRae, G.J. "Application of Market Mechanisms to Pollution." *Policy Studies Review* 1:3 (1982).

Hahn, R.W., and Noll, R.G. "Barriers to Implementing Tradable Air Pollution Permits: Problems of Regulatory Interactions." *Yale Journal on Regulation* 1:1 (1983).

Hamilton, M.M. "EPA Promotes Private Method of Air Cleanup." *Washington Post*, January 27, 1981.

Hardin, G. "The Tragedy of the Commons." *Science* 162 (December 13, 1968)

Highton, N.H., and Webb, M.G. "Sulphur Dioxide from Electricity Generation: Policy Options for Pollution Control." *Energy Policy*, March 1980.

Hoffman, J.S. "Economic Advantages of Emissions Banking Systems." Paper No. 79-54.7, presented at the 72nd Annual Meeting of the Air Pollution Control Association, Cincinnati, Ohio, June 1979.

———. "How to Reduce the Costs of Achieving and Maintaining Air Quality Standards in Metropolitan Areas." Presented at the American Planning Association Annual Meeting, Baltimore, Maryland, October 1979.

———. "Brokering Options." In *Brokering Emission Reduction Credits: A Handbook*. EPA Regulatory Reform Staff, Washington, D.C. January 1981.

ICF, Inc. *Concept Paper: Emission Reduction Banking and Trading Systems*. OPA 113/0, Washington, D.C., June 1980.

———. *Emission Reduction Banking Manual*. Washington, D.C., September 1980.

———. "Private Trading of Emission Reduction Credits." In *Brokering Emission Reduction Credits: A Handbook*. EPA Regulatory Reform Staff, Washington, D.C., January 1981.

———. "Alternative Strategies for Reducing Utility SO_2 and NO_x Emissions." Prepared for Argonne National Labs., Department of Energy and EPA, September 1981 (unpublished).

———. *Preliminary Analysis of the Mitchell Bill*. Washington, D.C.: Edison Electric Institute, March 1, 1982.

———. "Analysis of a Senate Emission Reduction Bill (S-3041)." Prepared for EPA Office of Policy Analysis, Office of Policy and Resource Management, February 1983 (unpublished).

Illinois Department of Energy and Natural Resources. *Illinois Energy Plan, Vol. 1: An Overview of Trends in the Illinois Electric Utility Industry*. ILLDOE 82/04, Springfield, Ill., July 1982.

Jablonka, G.E.; Endres, V.N.; and Neumeyer, D.G. "SO_2 Emission Reduction Optimization Program (SEROP)." Presented at the 46th Edison Electric Institute System Planning Meeting, Milwaukee, Wisconsin, May 17, 1982.

———. "Computer Programs Optimize SO_2 Emission Reduction Plans." *Power Engineering*, December 1982.

Jaedicke, C.F. "Strategic Planning: A View from the Bottom Up." *Public Utilities Fortnightly*, December 3, 1981.

Kelman, S. "Economists and the Environmental Muddle." *The Public Interest* 64 (Summer 1981).

Kerr, R.A. "Tracing Sources of Acid Rain Causes Big Stir." *Science* 215:4534 (February 12, 1982).

Kneese, A.V., and Schultze, C.L. *Pollution, Prices and Public Policy*. Washington, D.C.: The Brookings Institution, 1975.

Knudson, D.A., and Streets, D.G. "Mitigation of Acid Rain—Policy Alternatives." *Environmental Progress* 1:2 (May 1982).

Kontnik, L.T. "The Evolving Emission Trading Program: Fish or Fowl?" *Environmental Analyst* 3:9 (August 1982).

Kostow, L.P., and Kowalczyk, J.F. "A Practical Emission Trading Program." *Journal of the Air Pollution Control Association* 33:10 (October 1983).

Krupnick, A.J.; Oates, W.E.; and Van De Verg, E. "On Marketable Air-Pollution Permits: The Case for a System of Pollution Offsets." *Journal of Environmental Economics and Management* 10 (1983).

Krupp, F.D. "New Environmentalism Factors in Economic Needs." *Wall Street Journal*, November 20, 1986.

Kurtzweg, J.A., and Griffin, C.N. "Economic Development and Air Quality: Complementary Goals for Local Governments." *Journal of the Air Pollution Control Association* 31:11 (November 1981).

Kurtzweg, J.A., and Nelson, C.J. "Clean Air and Economic Development: An Urban Initiative." *Journal of the Air Pollution Control Association* 30:11 (November 1980).

Lakhani, H. "Air Pollution Control by Economic Incentives in the U.S.: Policy, Problems and Progress." *Environmental Management* 6:1 (January 1982).

Landau, J.L. "Who Owns the Air? The Emission Offset Concept and Its Implications." *Environmental Law* 9 (Spring 1979).

Levin, M.H. "The Supreme Court's Bubble Decision: What It Means." *EPA Journal*, September 1984.

———. "Building a Better Bubble at EPA." *Regulation*, March/April 1985.

———. "The Clean Air Act Needs Sensible Emissions Trading." *The Environmental Forum* 4 (March 1986).

Likens, G.E.; Wright, R.F.; Galloway, J.N.; and Butler, T.J. "Acid Rain." *Scientific American* 241:4 (October 1979).

Liroff, R.A., *Air Pollution Offsets: Trading, Selling, and Banking*. Washington, D.C.: The Conservation Foundation, 1980.

———. "The Bubble: Will It Float Free . . . or Deflate?" *The Environmental Forum* 4 (March 1986).

———. *Reforming Air Pollution Regulation: The Toil and Trouble of EPA's Bubble*. Washington, D.C.: The Conservation Foundation, 1986.

Lovelock, J. *Gaia*. Oxford: Oxford University Press, 1979.

Lukey, M.; Reed, G.; and Hillis, B. "Market Analysis of Emission Reduction Credits." In *Brokering Emission Reduction Credits: A Handbook*. EPA Regulatory Reform Staff, Washington, D.C., January 1981.

Mackintosh, D.R. *The Economics of Airborne Emissions: The Case for an Air Rights Market*. New York: Praeger Publishers, 1973.

Majone, G. "Choice Among Policy Instruments for Pollution Control." *Policy Analysis* 2:4 (Fall 1976).

Males, R. "SURE Findings and Conclusions." *EPRI Journal* 7:6 (July/August 1982).

Malone, L.P. "Air Quality and Development: How They Can Complement Each Other." *Commentary*, October 1979.

Marcus, A.A.; Sommers, P.; and Morris, F.A. "Alternative Arrangements for Cost Effective Pollution Abatement: The Need for Implementation Analysis." *Policy Studies Review* 1:3 (1982).

Marcus, S.J. "A Wall Street for Pollution?" *Technology Review* 84:2 (November/December 1981).

Massachusetts State Acid Rain Program. *Emissions Trading Within Acid Deposition Control: An Evaluation of Program Implementation and Regulatory Issues*. Boston: Division of AQ Control, Massachusetts Department of Environmental Quality Engineering, October 1986.

McLean, R.A.N. "The Relative Contribution of Sulfuric and Nitric Acids in Acid Rain to the Acidification of the Ecosystem: Implications for Control Strategies." *Journal of the Air Pollution Control Association* 31:11 (November 1981).

Molburg, J.C., and Rubin, E.S. "Air Pollution Control Costs for Coal-to-Electricity Systems." *Journal of the Air Pollution Control Association* 33:5 (May 1983).

Montgomery, W.D. "Markets in Licenses and Efficient Pollution Control Programs." *Journal of Economic Theory* 5 (1972).

National Commission on Air Quality. *To Breath Clean Air*. Washington, D.C., March 1981.

National Economic Research Associates, Inc. (NERA). "A Report on the Results from the Edison Electric Institute Study of the Impact of the Senate Committee on the Environment and Public Works Bill on Acid Rain Legislation (S.768)." June 20, 1983.

National Research Council. *Acid Deposition: Atmospheric Processes in Eastern North America*. Washington, D.C: National Academy Press, 1983.

Navarro, P. "The Politics of Air Pollution." *The Public Interest* 59 (Spring 1980).

———. "The 1977 Clean Air Act Amendments: Energy, Environmental, Economic and Distributional Impacts." *Public Policy* 29:2 (Spring 1981).

Neumeyer, D.G. "Analyzing Emission Controls for Power Plants." *Pollution Engineering* 15:4 (April 1983).

Noll, R.G. "Implementing Marketable Emissions Permits." *American Economic Review*, May 1982.

Norco, J.E., and Raufer, R.K. "Experience with an Emissions Trading Program." Presented at the 6th Annual Energy-Sources Technology Conference and Exhibition, Houston, Texas, January 1983.

"NSPS Bubble Issues." Prepared for discussion at the National Air Pollution Control Techniques Advisory Committee Meeting, June 1982.

Nulty, P. "A Brave Experiment in Pollution Control." *Fortune*, February 12, 1979.

O'Brien, K.M., and Euser, B. "Solar Energy and the Search for Emissions Offsets." *Solar Law Reporter* 1:6 (1979–80).

Office of Technology Assessment. *The Regional Implications of Transported Air Pollutants: An Assessment of Acidic Deposition and Ozone* (Interim Draft), Washington, D.C., July 1982.

Ostrov, J. "Interboundary Stationary Source Pollution—Clean Air Act Section 126 and Beyond." *Columbia Journal of Environmental Law* 8:1 (1982).

Palmisano,J. "An Evaluation of Emissions Trading." Presented at the 76th Annual Meeting of the Air Pollution Control Association, Atlanta, Georgia, June 1983.

———. "Emission Reduction Credits: Key to Cogen in Dirty-Air Districts." *Cogeneration*, May–June 1985.

Pasztor, A. "Market Booms for 'Rights' to Pollute." *Wall Street Journal*, June 18, 1981.

Pennsylvania Electric Co. "Continuous Emission Offsets" (n.d.).

"PG&E Plans Unique Power Mix for 1980s." *Electrical World* 196:7 (July 1982).

Pigou, A.C. *The Economics of Welfare*, 4th ed. New York: Macmillan 1932.

Putta, S.N. "Extension of Bubble Policy for Expanding the Use of Coal by New York Utilities." Presented at the Governor's Conference on Expanding the Use of Coal in New York State, Albany, New York, May 21 and 22, 1981.

Rahn, K.A. "A Statement of New Evidence." *The Environmental Forum* 1:1 (May 1982).

Rapoport, R. "Trading Dollars for Dirty Air." *Science 86*, July/August 1986.

Raufer, R.K.; Bodmer, E.C.; and Jaksch, J.A. "Tax and Financial Implications of ERC Leasing by Electric Utilities in Illinois." Paper No. 86-84.3, presented at the 79th Annual Meeting of the Air Pollution Control Association, Minneapolis, Minnesota, June 1986.

Raufer, R.K., and Feldman, S.L. *Analysis of the Potential for ERC Leasing Transactions Under Various Acid Deposition Control Requirements*. Philadelphia: University of Pennsylvania Energy Center, for Regulatory Reform Staff, EPA, January 1984.

———. "Emissions Trading and What It May Mean for Acid Deposition Control." *Public Utilities Fortnightly* 114:4 (August 16, 1984).

Raufer, R.K.; Feldman, S.L.; and Jaksch, J.A. "Emissions Trading and Acid Deposition Control: The Need for ERC Leasing." *Journal of the Air Pollution Control Association* 36:5 (May 1986).

Raufer, R.K.; Hill, L.G.; and Samsa, M.E. "Emission Fees and TERA: An Evaluation of Policy Alternatives in the Twin Cities." *Journal of the Air Pollution Control Association* 31:8 (August 1981).

RCF, Inc. *Emission Reduction Banking Experience in the United States and Its Applicability to Illinois*. Illinois Department of Energy and Natural Resources Doc. No. 84/04, March 1984.

Renshaw, E.F. "A Tax Subsidy Approach to Acid Rain Abatement." *Public Utilities Fortnightly*, September 30, 1982.

Reynolds, J.; Kreidenweis, S.; and Theodore, L. "Results of a Baghouse Operation and Maintenance Survey in Industry and Utility Coal-Fired Boilers." *Journal of the Air Pollution Control Association* 33:4 (April 1983).

Rittenhouse, R.C. "Environmentally Speaking." *Power Engineering* 86:4 (April 1982).

———. "Upgrading and Maintaining Coal Handling Equipment." *Power Engineering* 87:3 (March 1983).

Roberts, M.J., and Bluhm, J.S. *The Choices of Power: Utilities Face the Environmental Challenge*. Cambridge: Harvard University Press, 1981.

Roe, D. *Dynamos and Virgins*. New York: Random House, 1984.

Rogers, Golden, and Halpern, Inc. *New Jersey Energy Facility Development Potential Study*. Prepared for the New Jersey Department of Environmental Protection, Trenton, September 1981.

Rosencrantz, A. "Economic Approaches to Air Pollution Control." *Environment* 23:8 (October 1981).

Rubin, E.S.; Cushey, M.A.; Marnicio, R.J.; Bloyd, C.N.; and Skea, J.F. "Controlling Acid Deposition: The Role of FGD." *Environmental Science and Technology* 20:10 (October 1986).

"Rules Establishing Pollution Charges and Restating Permit Application Fees in Accord with Title 10, Vermont Statutes Annotated, Chapter 33, as Amended." Resources Board, State of Vermont, June 29, 1979.

Rupp, S. "One Broker's Experience." In *Brokering Emission Reduction Credits: A Handbook*. EPA Regulatory Reform Staff, Washington, D.C., January 1981.

Russell, C.S. "What Can We Get from Effluent Charges?" *Policy Analysis* 5:2 (Spring 1979).

———. "Controlled Trading of Pollution Permits." *Environmental Science and Technology* 15:24 (January 1981).

Sandbach, G.F. *Environment, Ideology and Policy*. Oxford: Basil Blackwell, 1980: Totowa, New Jersey: Rowman & Allanheld, 1980.

Schmitz, T.M. "Pollution, Law, Science and Damage Awards." *Cleveland Law Review* 18:3 (September 1969).

Schneider, A.M. "An Effluent Fee Schedule for Air Pollutants Based on PINDEX." *Journal of the Air Pollution Control Association* 23:486 (1973).

Schultze, C. *The Public Use of Private Interest*. Washington, D.C.: The Brookings Institution, 1977.

Science and Policy Associates. "Acidic Deposition Catalog." Prepared for EPRI, August 1986.

"Selling the Right to Pollute." *Washington Post*, August 18, 1980.

Seltz-Petrach, A. "Marketplace Solutions to Air Pollution." *Civil Engineering*, January 1980.

Seneca, J.J., and Tausig, M.K. *Environmental Economics*. Englewood Cliffs, New Jersey: Prentice-Hall, 1974.

Shabecoff, P. "U.S. Backs Trading in 'Credits' Allowing New Plants to Pollute." *New York Times*, January 26, 1981.

Silverman, B.G. "Acid Rain and a Comparative Analysis of Two Large Scale Electric Utility Models." *Journal of the Air Pollution Control Association* 31:10 (October 1982).

Stathos, D.T., and Treitman, M.S. "Using Private Market Incentives for Air Cleanup." *Public Utilities Fortnightly*, July 30, 1981.

Streets, D.G.; Conley, L.A.; Carter, L.D.; and Vernet, J.E. *An Analysis of Proposed Legislation to Control Acid Rain*. Argonne National Laboratory, ANL/EES-TM-209, January 1983.

Streets, D.G.; Garvey, D.B.; Grogan, P.J.; Hanson, D.A.; and Carter, L.D. "A Regional, New Source Bubble Policy: Its Advantages Illustrated for the State of Illinois." *Journal of the Air Pollution Control Association* 34:1 (January 1984).

Streets, D.G.; Knudson, D.A.; and Shannon, J.D. "Selected Strategies to Reduce Acidic Deposition in the U.S." *Environmental Science and Technology* 17:10 (1983).

Stretton, H. *Capitalism, Socialism and the Environment*. Cambridge: Cambridge University Press, 1976.

Teknekron Research, Inc. "Electric Utility Emissions: Control Strategies and Costs." R-005-EPA-81/R1, Prepared for EPA, Argonne National Labs and the U.S. Department of Energy, April 1981.

Tietenberg, T.H. "Transferable Discharge Permits and the Control of Stationary Source Air Pollution: A Survey and Synthesis." *Land Economics* 56:4 (November 1980).

———. *Emissions Trading: An Exercise in Reforming Pollution Policy*. Washington, D.C.: Resources for the Future, 1985.

Tobin R.J. "Is Regulation Outdated? Achieving Clean Air with Alternatives to Regulation." Presented at the Midwest Political Science Association Meeting, Cincinnati, Ohio, April 1981.

Trisko, E.M. "Acid Precipitation: Causes, Consequences, Controls," *Public Utilities Fortnightly*, February 3, 1983.

Trisko, E.M., and Wayland, R.E. "Acid Rain Control and Public Utility Regulation." *Public Utilities Fortnightly*, August 30, 1984.

Tucker, W. "Marketing Pollution." *Harper's* 262:1572 (May 1, 1981).

U.S. Council on Environmental Quality. *Annual Report on Environmental Quality for 1984*. Table A-19, "National Expenditures for Pollution Abatement and Control, 1972–1983," 1985.

U.S. Department of Energy. *Inventory of Power Plants in the United States: 1981 Annual*. Energy Information Administration, DOE/EIA-0095 (81), September 1982.

———. *Inventory of Power Plants in the United States: 1982 Annual*. Energy Information Administration, DOE/EIA-0095 (82), June 1983.

U.S. Regulatory Council. *Alternative Regulatory Approaches: An Overview*. Project on Alternative Regulatory Approaches, Washington, D.C., September 1981.

———. *Marketable Rights: A Practical Guide to the Use of Marketable Rights as a Regulatory Alternative*. Project on Alternative Regulatory Approaches, Washington, D.C., September 1981.

"Used Smog for Chi-Town." *Mother Jones*, November, 1979, p. 12.

"Utility's New SEROP Model Provides Optimized Emission Reduction Strategies." *Journal of the Air Pollution Control Association* 32:9 (September 1982).

Vatavuk, W.M., and Neveril, R.B. "Estimating Costs of Air Pollution Control Systems, Part XI: Estimate the Size and Cost of Baghouses." *Chemical Engineering*, March 22, 1982.

Violette, D.M. "Comment on Decision Problems in the Control of Acid Precipitation: Nonconvexities and Irreversibilities." *Journal of the Air Pollution Control Association* 34:1 (January 1982).

Vivian, W., and Hall, W. *An Examination of the U.S. Market Trading in Air Pollution Offsets*. Ann Arbor: University of Michigan Institute of Public Policy Studies, March 1981.

Watson, S.A. "Banking and the Emission Offset Policy: A Market Approach to Pollution Control." Unpublished manuscript, University of California, Berkeley, June 1979.

Wenner, L.M. *One Environment Under Law*. Pacific Palisades, California: Goodyear Publishing Co., 1976.

———. "Pollution Control: Implementation Alternatives." *Policy Analysis* 4 (Winter 1978).

Wetstone, E.G. "The Need for a New Regulatory Approach." *Environment* 22:5 (June 1980).

Wildavsky, A. *The Politics of the Budgetary Process*, 3rd ed. Boston: Little, Brown, 1979.

———. *Speaking Truth to Power*. Boston: Little, Brown, 1979.

Willingham, M.G., *Analysis of Select Gas Use, Emission Reduction Credits and Limestone Injection Controls in Acid Deposition Control Programs*. Prepared for Regulatory Reform Staff, EPA, February 1986.

World Resources Institute (WRI). *World Resources 1986*. New York: Basic Books, 1986.

Yandle, B. "The Emerging Market in Air Pollution Rights." *Regulation*, July/August 1978.

Zanoni, R.A., and Robison, A.J. "Leasing—A Preferred Alternative for Utilities?" *Public Utilities Fortnightly*, July 11, 1985.

Index

F

G

H

I

J

K

L

M

N

O

P